# Conceptual Metaphor and Embodied Cognition in Science Learning

Scientific concepts are abstract human constructions, invented to make sense of complex natural phenomena. Scientists use specialised languages, diagrams, and mathematical representations of various kinds to convey these abstract constructions. This book uses the perspectives of embodied cognition and conceptual metaphor to explore how learners make sense of these concepts. That is, it is assumed that human cognition – including scientific cognition – is grounded in the body and in the material and social contexts in which it is embedded. Understanding abstract concepts is therefore grounded, via metaphor, in knowledge derived from sensory and motor experiences arising from interaction with the physical world.

The volume consists of nine chapters that examine a number of intertwined themes: how systematic metaphorical mappings are implicit in scientific language, diagrams, mathematical representations, and the gestures used by scientists; how scientific modelling relies fundamentally on metaphor and can be seen as a form of narrative cognition; how implicit metaphors can be the sources of learner misconceptions; how conceptual change and the acquisition of scientific expertise involve learning to coordinate the use of multiple implicit metaphors; and how effective instruction can build on recognising the embodied nature of scientific cognition and the role of metaphor in scientific thought and learning. The volume also includes three extended commentaries from leading researchers in the fields of cognitive linguistics, the learning sciences, and science education, in which they reflect on theoretical, methodological and pedagogical issues raised in the book.

This book was originally published as a special issue of the *International Journal of Science Education*.

**Tamer G. Amin** is Associate Professor at the Science and Mathematics Education Center in the Department of Education at the American University of Beirut, Lebanon. His research focuses on the development of scientific understanding and reasoning, with a particular emphasis on how language relates to this process. He is currently co-editing (with Olivia Levrini) a forthcoming Routledge publication entitled *Converging Perspectives on Conceptual Change: Mapping an Emerging Paradigm in the Learning Sciences*.

**Fredrik Jeppsson** is Assistant Professor in the Department of Social and Welfare Studies at Linköping University, Sweden. His research focuses on the teaching and learning of thermal science at different age levels. He has an interest in embodied cognition and, in particular, how metaphorical language can support students' learning. He currently studies how primary students' meaning-making of science content can be promoted by mediation through different semiotic modes (e.g. verbal language, body language, images, or models).

**Jesper Haglund** is a Docent in Physics with specialisation in Physics Education at the Department of Physics and Astronomy, Uppsala University, Sweden. His research focuses on the teaching and learning of thermal science at different age levels. He has a particular interest in embodied cognition and how analogies and metaphors can support students' learning. Recently, his study has focused on how infrared cameras can be used in open-ended laboratory activities.

# Conceptual Metaphor and Embodied Cognition in Science Learning

*Edited by*
Tamer G. Amin, Fredrik Jeppsson and Jesper Haglund

LONDON AND NEW YORK

First published 2017
by Routledge

2 Park Square, Milton Park, Abingdon, Oxfordshire OX14 4RN
52 Vanderbilt Avenue, New York, NY 10017

*Routledge is an imprint of the Taylor & Francis Group, an informa business*

First issued in paperback 2018

Introduction, Chapters 1-4, 6-11 © 2017 Taylor & Francis
Chapter 5 © Hunter G. Close and Rachel E. Scherr

*British Library Cataloguing in Publication Data*
A catalogue record for this book is available from the British Library

ISBN 13: 978-1-138-23075-0 (hbk)
ISBN 13: 978-0-367-07584-2 (pbk)

Typeset in Plantin
by RefineCatch Limited, Bungay, Suffolk

**Publisher's Note**
The publisher accepts responsibility for any inconsistencies that may have
arisen during the conversion of this book from journal articles to book chapters,
namely the possible inclusion of journal terminology.

**Disclaimer**
Every effort has been made to contact copyright holders for their permission to
reprint material in this book. The publishers would be grateful to hear from any
copyright holder who is not here acknowledged and will undertake to rectify
any errors or omissions in future editions of this book.

# Contents

# CONTENTS

# Citation Information

The chapters in this book were originally published in the *International Journal of Science Education*, volume 37, issues 5–6 (March 2015). When citing this material, please use the original page numbering for each article, as follows:

## Introduction

*Conceptual Metaphor and Embodied Cognition in Science Learning: Introduction to special issue*
Tamer G. Amin, Fredrik Jeppsson and Jesper Haglund
*International Journal of Science Education*, volume 37, issues 5–6 (March 2015)
pp. 745–758

## Chapter 1

*The Importance of Language in Students' Reasoning About Heat in Thermodynamic Processes*
David T. Brookes and Eugenia Etkina
*International Journal of Science Education*, volume 37, issues 5–6 (March 2015)
pp. 759–779

## Chapter 2

*Varying Use of Conceptual Metaphors across Levels of Expertise in Thermodynamics*
Fredrik Jeppsson, Jesper Haglund and Tamer G. Amin
*International Journal of Science Education*, volume 37, issues 5–6 (March 2015)
pp. 780–805

## Chapter 3

*On Conceptual Metaphor and the Flora and Fauna of Mind: Commentary on Brookes and Etkina; and Jeppsson, Haglund, and Amin*
Bruce Sherin
*International Journal of Science Education*, volume 37, issues 5–6 (March 2015)
pp. 806–811

**Chapter 11**
*Conceptual Metaphor and the Study of Conceptual Change: Research synthesis and future directions*
Tamer G. Amin
*International Journal of Science Education*, volume 37, issues 5–6 (March 2015)
pp. 966–991

For any permission-related enquiries please visit:
http://www.tandfonline.com/page/help/permissions

# Notes on Contributors

**Tamer G. Amin** is Associate Professor at the Science and Mathematics Education Center in the Department of Education at the American University of Beirut, Lebanon.

**David T. Brookes** is Assistant Professor of Physics in the Physics Education Research Group at Florida International University, Miami, USA.

**Hunter G. Close** is Associate Professor of Physics at Texas State University, San Marcos, Texas, USA.

**Benjamin W. Dreyfus** is Research Associate in the Department of Physics at the University of Maryland, College Park, MD, USA.

**Reinders Duit** is Professor in the Leibniz Institute for Science Education, University of Kiel, Germany.

**Eugenia Etkina** is a Professor working in the field of Physics Education Research in the Graduate School of Education at Rutgers University, NJ, USA.

**Hans U. Fuchs** is Professor of Physics in the Institute of Applied Mathematics and Physics at Zurich University of Applied Sciences at Winterthur, Switzerland.

**Harald Gropengiesser** is a Professor in the Institute for Science Education at Leibniz University Hannover, Germany.

**Ayush Gupta** is Research Assistant Professor in the Department of Physics at the University of Maryland, College Park, MD, USA.

**Jesper Haglund** is a Docent in Physics with specialisation in Physics Education at the Department of Physics and Astronomy, Uppsala University, Sweden.

**Fredrik Jeppsson** is Assistant Professor in the Department of Social and Welfare Studies at Linköping University, Sweden.

**Rachael Lancor** is Assistant Professor of Physics at Edgewood College, Madison, WI, USA.

**Kai Niebert** is Professor of Science Education and Sustainability at the University of Zurich, Switzerland.

**Rafael Núñez** is Professor in the Department of Cognitive Science at the University of California – San Diego, CA, USA.

**Edward F. Redish** is Professor of Physics at the University of Maryland, College Park, MD, USA.

**Rachel E. Sherr** is Senior Research Scientist in the Physics Department at Seattle Pacific University, Seattle, WA, USA.

**Bruce Sherin** is Professor of Learning Sciences and Co-ordinator of the Learning Sciences PhD programme at Northwestern University, Evanston, IL, USA.

**David F. Treagust** is Professor of Science Education in the Science and Mathematics Education Centre at Curtin University, Perth, Australia.

# Conceptual Metaphor and Embodied Cognition in Science Learning: Introduction

Tamer G. Amin[a], Fredrik Jeppsson[b] and Jesper Haglund[c]

[a]Science and Mathematics Education Center, Department of Education, American University of Beirut, Beirut, Lebanon; [b]Department of Social and Welfare Studies, Linköping University, Norrköping, Sweden; [c]Department of Physics and Astronomy, Uppsala University, Uppsala, Sweden

## Orientation and Theoretical Background

We introduce here a special issue of this journal on the theme of 'Conceptual Metaphor and Embodied Cognition in Science Learning'. The idea for this issue grew out of a symposium that we organized on this topic at the conference of the European Science Education Research Association (ESERA) in September 2013. The eight papers collected in this issue reflect the emergence of a critical mass of studies in science education applying ideas from the perspective of 'embodied cognition' in cognitive science. Up until the 1980s, most research in cognitive science assumed a view of the mind as an abstract information processing system. On this view, our sensorimotor systems were often seen as serving a peripheral, input/output role, conveying information to or from a central cognitive processor where abstract, higher level thought took place. The research focused on developing models of cognition incorporating language-like, propositional representations and syntactic processes, and largely ignored the specifics of human physiology and interaction between the person and the material and social world in which he or she thinks and acts. Since then, several different approaches to cognitive science have adopted some version of the assumption that cognition is embodied—that is, they have assumed that models of cognition need to attend to the characteristics of human brains and bodies, and the material contexts in which thought is taking place (e.g. Barsalou,

2008; Clark & Chalmers, 1998; Shapiro, 2011; Varela, Thompson, & Rosch, 1991; Wilson, 2002). The broad assumptions behind embodied cognition are not new to the study of the mind and may be traced back to Merleau-Ponty's (1962/2002) *Phenomenology of perception* and Gibson's (1979) ecological theory of perception. They are also acknowledged in cognitive developmental traditions, such as the Piagetian emphasis on our sensorimotor system as a basis for the development of abstract concepts, and resonate with Vygotsky's (1978) recognition of the role of our interaction with physical and symbolic artifacts. With regard to the educational sciences, certain ideas of embodied cognition are in line with pragmatic and progressive traditions, for example, those of Dewey (1916) which emphasize the role of personal and physical experiences in learning.

Wilson (2002) carefully distinguishes and assesses six distinct claims that fall under the general heading of embodied cognition: (1) that cognitive processes are situated, varying depending on the real-world contexts in which they are carried out; (2) that cognitive processes must be understood with respect to the specific temporal constraints imposed on our brains by the environment when cognitive tasks are carried out; (3) that cognitive processes recruit the material, symbolic and social structure of the environment, reducing what actually needs to be performed in the mind itself; (4) that cognitive systems can be viewed as extended, where there is no sharp divide between internal and external contributions to cognition; (5) that the function of cognition is not primarily to represent the external world but to guide action in it and (6) that even cognition that takes place in the 'mind' proper relies on knowledge structures that emerge from body-based experiences. This introduction is not the place for a discussion of Wilson's evaluation of these claims. We simply note that she finds the fourth claim 'deeply problematic' but cautiously accepts the first three and fifth claims, suggesting that the range of applicability of each still needs to be more fully assessed. The sixth claim she considers to be the most powerful of all the claims and reviews evidence suggesting that body-based cognitive representations and processes ground a wide range of 'off-line' mental phenomena such as mental imagery, working memory, episodic memory, implicit memory, and reasoning and problem solving. The research included in this special issue relates to the third and sixth claims reviewed by Wilson. We elaborate on each of these two claims before introducing the eight papers and three commentaries included in the issue.

We begin with the claim (Wilson's Claim 6) that even cognition that takes place in the 'mind' proper relies on knowledge structures that emerge from body-based experiences. Research in neuroscience, cognitive psychology and cognitive linguistics has found that mental imagery relies on the same brain mechanisms used for perception and action, that the multicomponent working memory system includes a component that stores visuospatial information to be used to carry out cognitive tasks, and that abstract concepts are understood in terms of generalizations over sensorimotor experiences ('image schemas') via metaphorical mapping, with reasoning using these abstract concepts also relying on the inferences generated by image schemas (Gibbs, 2005). The latter claim, developed under the label 'conceptual metaphor theory' within the field cognitive linguistics has been very influential in the research represented in this special issue.

Conceptual metaphor theory was originally developed by Lakoff and Johnson (1980, 1999). They argued that our conceptual system develops through personal, physical experiences as we interact with the surrounding world. At the most basic level, we form *image schemas*, knowledge gestalts that emerge out of repeated sensor-imotor experiences when interacting with the surrounding world (Johnson, 1987). Examples of image schemas include the *container schema*, in which we conceptualize an inside, an outside and a separating boundary; and the *source-path-goal schema*, through which we conceptualize an object moving along a path, from a source to a goal. Lakoff and Johnson suggested that these image schemas ground our understand-ing of abstract concepts and our use of language. One of their central claims is that image schemas may be mapped metaphorically to more abstract domains forming *conceptual metaphors*. For instance, by mapping the source-path-goal schema onto the abstract concept of 'love', we may form the conceptual metaphor LOVE IS A JOURNEY, an underlying cognitive structure that finds expression in utterances such as 'We're at a crossroads' or 'We may have to go our separate ways'.

Conceptual metaphor theory was developed initially based on the identification of pervasive and systematic patterns in metaphorical expressions that were found to reflect mappings between conceptual domains of knowledge. Follow-up research in psycholinguistics, neuroscience and gesture analysis has provided further evidence of the psychological reality of the role of image schemas in grounding interpretation of language and reasoning (Gallese & Lakoff, 2005; Gentner, Bowdle, Wolff, & Boronat, 2001; Gibbs, 2005; Núñez & Sweetser, 2006). As we describe below, the contributors to this special issue have used the analysis of language and gesture as methods to identify image schemas invoked by students, teachers and scientists and how they map them metaphorically onto abstract scientific concepts they are thinking about and reasoning with.

The second claim from embodied cognition that features prominently in the research included in this issue (Wilson's Claim 3) is that people recruit the material, symbolic and social structure of their environment, reducing what they actually need to perform in the mind itself. The idea here is that highly complex and abstract cog-nitive tasks can be guided, simplified or off-loaded onto the structure of physical objects; symbolic representations such as diagrams, mathematical equations and language; and the actual organization of social structures. A prominent example in the embodied cognition literature is Hutchins' (1995) analysis of the complex cogni-tive task of navigating a large naval vessel. He shows how the cognitive processes taking place within each individual involved in the process is simplified by the actual objects in the environment, the instruments and charts used to identify the vessel's location, and the social hierarchy of the officers on the vessel. Another phenomenon that illustrates this claim, in the case of language in particular, has been identified through another strand of research in cognitive linguistics—namely, research on 'conceptual integration' or 'blending' developed by Fauconnier and Turner (1998). Also accepting embodied cognition assumptions, this line of research has focused on describing the process of meaning construction that takes place as discourse unfolds. From this perspective, constructing meaning involves invoking and integrating conceptual content, triggered by linguistic forms (or elements of other symbol systems). Central to this model is the idea that conceptual content comes from distinct conceptual domains (or 'spaces'). Analogical mapping is one well-known example of this process. However, Fauconnier and Turner show that integrating conceptual content while using language can often go beyond mapping from one domain to another. It can often involve *blending* conceptual content from more than one domain. The outcome of this process is often to greatly simplify the interpretation of complex and abstract meanings and reasoning sequences. Many mundane, and not so mundane, uses of language reveal many subtle and varied mappings once subjected to a conceptual integration analysis.

Let us illustrate how this framework is applied using an example from Turner and Fauconnier (1995). A catamaran sailed from San Francisco to Boston in 1993 in an attempt to break the record established by a clipper in 1853. At some stage after the journey began, a newspaper reported that 'the catamaran was "barely maintaining a 4.5 day lead" over the clipper' (cited in Turner & Fauconnier, 1995). What could 'maintaining a lead' mean here? Turner and Fauconnier suggest that the phrase is understood as referring to a highly simplified fictitious situation where both the catamaran and the clipper are sailing from San Francisco to Boston at the same time. They analyze the interpretation of the phrase in terms of mapping between four conceptual 'spaces'. They argue that the key to interpreting the phrase is the construction by the listener of a 'blended space'. They explain that there are two input conceptual spaces (one for each of the 1993 and 1853 sails), a generic space that is structured internally with an abstract schema (e.g. some boat sailing between two cities at some unspecified time), and a blend that is structured by partial input from the two input spaces and the generic space. Both the catamaran and clipper are projected into the blend along with many specifics of the journey of each. However, the specific dates are not projected to the blend because including two different times in a single conceptual space would be internally inconsistent. Only a generic time is projected from the generic space to the blend, establishing the idea that the two sails are occurring at the same time. They also point out that what is crucial for the reader to understand the newspaper report is that the fictitious blended space incorporates conceptual relations that are absent in either of the input spaces alone. The presence of two boats simultaneously on the path between the two cities involves relative position between the boats absent in either input. This then encourages a person making sense of the phrase to invoke an imaginary race frame, with a winner and a loser, a sense of competition, etc. Of particular importance for the research included in this special issue is that the blend simplifies the conceptual processes needed to compare the two sails at disparate points in time, compressing it into a race frame. Fauconnier and Turner (2002) have discussed how such blends are pervasive in language as well as in the construction and interpretation of other symbol systems like gestures, diagrams and objects with symbolic meaning (e.g. a watch face). As will be explained below, research in science education has begun to explore how language, gesture and objects (including the human body) can be shown to support the simplification of scientific conceptualization and reasoning in abstract domains.

The science education research included in this special issue applies these two assumptions from embodied cognition—the grounding of mental processes in body-based knowledge structures and the offloading and simplification of cognitive processes onto external objects and symbols (including both language and gesture). This research uses theoretical constructs and methods from the two strands of research in cognitive linguistics introduced above: conceptual metaphor theory (Lakoff & Johnson, 1980, 1999) and conceptual integration or blending theory (Fauconnier & Turner, 1998). This research in science education is not unique in the educational literature. Glenberg (2008) and Kontra, Goldin-Meadow, and Beilock (2012) have recognized the role of embodied cognition in education and concept learning. In particular, a special issue was recently dedicated to embodied approaches to mathematics education in the *Journal of the Learning Sciences* (Hall & Nemirovsky, 2012). The perspectives of conceptual metaphor and conceptual integration have been used to analyze language use and conceptualization in technical, as well as everyday thought and language. Lakoff and Núñez (2000) show that conceptual metaphors are heavily involved in how we structure one of the most formal and abstract areas of human thought—mathematics. For instance, the idea of an equation builds on a balance schema, involving experiences such as equal weights on a scale. Similarly, building on conceptual metaphors, Sfard (1994) shows how we treat abstract entities, such as numbers and functions, as objects through a process of reification. The special issue on embodied learning in mathematics, referred to above, involves themes such as how we make use of the *container*, *proximity* and *source-path-goal* schemata in learning to read analog clocks (Williams, 2012), how learning of complex numbers is enhanced by body motion on a tiled floor (Nemirovsky, Rasmussen, Sweeney, & Wawro, 2012), and how gestures can be used to reveal image-schematic underpinnings of mathematical cognition (Alibali & Nathan, 2012).

Within science education, the ideas of Lakoff and Johnson (1980) were possibly first recognized by Andersson (1986) who suggested that learners make sense of a range of phenomena, including phase transitions, mechanics and electric circuits, by use of an *experiential gestalt of causation*, a schema of how an agent affects an object in interaction. Within physics education research, Podolefsky and Finkelstein (2007) have analyzed how increasingly abstract phenomena have been interpreted in terms of waves (from water waves, through sound and electromagnetism, to quantum phenomena) from the perspective of blending theory. Wittmann (2010) has used blending theory to explain that it is tempting, although misleadingly, for students to use their experience of throwing balls when identifying factors that impact the speed of propagation of a wave on a suspended string. However, questions of embodiment and the use of theories from cognitive linguistics have only recently received sustained and systematic attention in science education. The eight contributions to this special issue represent a number of research programs central to this body of work. All contributions draw on assumptions from the perspective of embodied cognition to examine issues in scientific cognition, science learning and instruction. Collectively, the contributions address concepts that span the areas of physics, chemistry, biology and climatology. As we point out below, the papers engage with issues central to research in science education such as the difference between novice and expert thinking, including differences in how concepts are categorized ontologically; the nature and source of student conceptions; the role of metaphor and analogy in concept learning and the role of representations and narrative in science instruction.

## Contributions to this Special Issue

The issue includes eight papers and three commentaries. The papers are grouped thematically with each of three sets of papers followed by a commentary. The first two papers (by Brookes and Etkina; and Jeppsson, Haglund and Amin) address the issue of how the phenomenon of conceptual metaphor figures in the acquisition of scientific expertise. In a commentary, Bruce Sherin discusses the basic theoretical assumptions guiding these two studies and how they can be situated in relation to other work in science education. The next set of two papers (by Dreyfus, Gupta and Redish; and Close and Scherr) uses the perspective of conceptual integration (or blending). These papers explore how learners and scientists blend multiple metaphors and how thinking with conceptual metaphors interacts with the use of external representations. These papers use analyses of language and gesture to argue that such blends and interactions are productive in learning and suggest approaches to instruction of difficult concepts. In his commentary on this second set, Rafael Núñez engages in careful analysis of the methods used in these two studies, focusing in particular on the analysis of gestures. A third set of three papers (by Lancor; Niebert and Gropengießer; and Fuchs) addresses the contribution of a conceptual metaphor perspective to identifying the narrative structure inherent in science, analyzing student and scientist conceptions, and designing instructional representations. In a joint commentary, David Treagust and Reinders Duit reflect on these three papers and examine the extent to which this work goes beyond prior research on narrative, metaphor, analogy and multiple representations in science education. The eighth paper in the issue (by Amin) reviews the literature on conceptual metaphor in science education to identify its contributions to the study of conceptual change and suggests directions for future research. In the remainder of this section, we introduce each of these contributions, situating them very briefly in relation to the authors' programs of research.

Brookes and Etkina (2007, 2009) began a program of research in the last few years, drawing on the theory of conceptual metaphor to analyze the language of science (in the domains of force and motion, and quantum mechanics), to describe the ontological classification of concepts (Chi & Slotta, 1993) implicit in this language, and to reveal conceptual misunderstandings of students that can be traced to these implicit ontologies. Their contribution in this special issue continues this line of work. However, in their paper here, they address the connections between language and students' conceptual difficulties more directly and make a clear distinction between *explicit* ontological beliefs about a concept (in this case heat) and conceptualizations *implicit* in language but revealed through conceptual metaphor analysis. Their study provides evidence linking non-canonical explicit ontological beliefs about the meaning of the word 'heat' and conceptualizations of heat as a substance implicit in students' language, on the one hand, to incorrect reasoning with the concept of heat as if it is a state function when solving thermodynamics problems, on the other. While providing empirical evidence for this link between language and state-function reasoning about heat, Brookes and Etkina are cautious not to impose a simple causal interpretation. Instead, they assume a bidirectional causal relationship between the conceptualizing of experience and language. From this perspective on language and conceptualization in science and based on their empirical results, they articulate an approach to science instruction as guided meaning making (following Lemke, 1998). In this approach, students are encouraged to avoid using technical terminology early on as they make sense of observations and construct explanations of phenomena. As a technical term is introduced, students are expected to make sure that its use and the meaning attributed to it are shared in the classroom. Brookes and Etkina believe that they can, thereby, avoid both extremes of allowing students to unproductively use language with misleading implicit ontologies or the forced emphasis on using ontologically 'precise' language advocated by some science educators.

In the next contribution to this issue, Jeppsson, Haglund and Amin hypothesize that implicit metaphorical construals of concepts such as heat and entropy (which are 'incorrect' from a scientific perspective) can contribute productively to expert scientific reasoning. In previous work together with Strömdahl, they have shown the pervasive and systematic use of such metaphorical construals of the concept of entropy and the second law of thermodynamics in university level textbooks (Amin, Jeppsson, Haglund, & Strömdahl, 2012) and scientific problem solving carried out by Ph.D. students (Jeppsson, Haglund, Amin, & Strömdahl, 2013). That work showed that substance-like construals of abstract concepts seem to be productively used by experts in problem solving and in communicating ideas to learners. The work reported by Jeppsson et al. here, contrasts the use of conceptual metaphors in problem solving at two levels of expertise: Ph.D. students and undergraduates. A pair of undergraduate students were given the same thermodynamics problems involving the concept of entropy that were given to the pair of Ph.D. students in Jeppsson et al. (2013). Qualitative analysis of the problem solving protocol for each of the pair of students revealed differences in how the two pairs used conceptual metaphors. The authors distinguish their approach to others in the expertise literature who focus either on the role of propositional representations (e.g. Chi, Feltovich, & Glaser, 1981) or on the role of non-propositional representations or processes (such as imagery, mental models or analogical reasoning) (e.g. Clement, 2009). In contrast, Jeppsson et al. interpret the patterns of use of the conceptual metaphors at different levels of expertise in terms of the nature of the coordination between propositional and non-propositional knowledge resources and processes. A key hypothesis from the analysis was that the more expert problem solvers (the Ph.D. students) used conceptual metaphors more extensively and in a less conventional way than the undergraduates, while constraining their use of these metaphors in light of propositional laws and principles that were invoked initially to launch the problem solving. Jeppsson et al. argue that the strategic use of conceptual metaphor in coordination with propositional principles is a feature of problem solving that needs to be acquired with expertise. Given the highly implicit nature of this aspect of problem solving, they argue that exposure to how experts coordinate these resources in apprenticeship settings must be an important component of instructional environments.

Bruce Sherin has contributed a commentary on the first two papers in this issue. He engages in a broad discussion of the theoretical frameworks adopted in these two papers and how they position themselves with respect to other lines of work in science education. He comments that he views the major contributions of these papers to be their attention to the more advanced levels of scientific expertise, to how a variety of different resources are weaved together in advanced scientific thought, and the attention given to language a tool for thought, not just as a window *onto* thought of value to the researcher. Sherin, however, takes issue with how Jeppsson, Haglund and Amin position their research with respect to other related work in science education. In addition, he argues that the construct of conceptual metaphor is used without attention to nuances among different kinds of mental structures. He also rejects what he sees as an exclusive assumption that concepts can only derive meaning by grounding in the body and insists that a notion of mental representation of concepts independent of external representations like language is needed. In sum, he argues for a more inclusive approach to the 'flora and fauna of the mind' when researching science learning and instruction.

The next two articles (by Dreyfus, Gupta and Redish; and Close and Scherr) use the blending framework to study scientists', teachers' and students' understanding and representation of the concept of energy, by analyzing language, gesture and other embodied activities. In the first of this pair, Dreyfus et al. build on their research on context-dependence and flexibility of ontological categorization in science and science learning (Gupta, Hammer, & Redish, 2010) and ontological metaphors for negative energy (Dreyfus et al., 2014). They examine how a physics professor and one of his students make use of a blend of two metaphors for energy when conceptualizing chemical bonds: energy-as-substance and energy-as-location. In the episodes they analyze, they find that the metaphor energy-as-substance is expressed in language, while energy-as-location is expressed simultaneously by means of gestures and the vertical dimension of a graph representing energy levels on the whiteboard. They argue that the two are integrated, forming a coherent blend. The blend is introduced by the professor in a lecture and later adopted by the student in a subsequent interview on the energy transfer involved in ATP synthesis. Dreyfus et al. argue that this finding is not consistent with the view that students' and scientists' conceptions can be classified into distinct ontological categories (Chi, Slotta, & De Leeuw, 1994). In contrast, they show that physicists and students may embrace, simultaneously, at least two such categories. This research shows how the substance metaphor of energy can be complemented with other construals of energy to give a more comprehensive idea of the concept in physics teaching.

The second paper in this pair reports on part of a project that looks to redesign undergraduate science teaching and conduct innovative professional development for teachers at Seattle Pacific University. In this Energy Project, Scherr and colleagues have developed representations and a learning environment that help learners adopt embodied construals of physical processes involving energy transfers and transformations. In their previous work, they explicitly exploited a substance metaphor for energy (Scherr, Close, McKagan, & Vokos, 2012). An example is the Energy Theater, a kind of structured, embodied role play, in which participants represent one unit of energy each with their bodies and depict the objects that contain the unit of energy through their location on a floor delimited with loops of rope and the form of energy with gestures (Scherr et al., 2013). In their contribution here, Close and Scherr analyze participants' Energy Theater enaction of a physical scenario, adiabatic compression of a gas, using blending theory. The physical scenario to be represented and the setting of the Energy Theater, with its ropes and the participants themselves, constitute the two input spaces for the blend. Certain characteristics of energy are anchored materially and socially (Hutchins, 1995) through the rules of Energy Theater. For instance, energy conservation is guaranteed as part of the game, since the participants themselves—corresponding to units of energy—cannot suddenly appear or disappear. The participants can therefore focus their attention on understanding the specific nature of physical scenarios they are confronted with. Close and Scherr analyze participants' discussions and bodily enactments using blending theory to show how the participants make conceptual leaps in their understanding of energy during an Energy Theater performance.

Rafael Núñez has contributed a commentary on these two closely related papers. Núñez is immersed in the foundational cognitive science literature on conceptual metaphor and blending, having contributed to its theoretical development (in particular in the domain of mathematical cognition) and to the development of methods, including the analysis of gesture. He brings this background to his discussion of these two papers. He is excited by the extension of these contemporary theories and methods in cognitive science to the context of science education, with the rich possibilities it brings to investigating the complexity of multimodal meaning making in science classrooms. He comments, however, that this complexity brings methodological challenges. He discusses two in particular, illustrating these through a close, critical reading of the analyses presented by Dreyfus et al. and Close and Scherr. The first problem he points out involves the characterization of the source domains of the conceptual metaphors identified. The second is the diversity of kinds of gestures and the challenge this diversity raises for inferring when particular gestures do or do not provide evidence for blending. He concludes by challenging science education researchers drawing on embodied cognition theories and methods to fine tune their methods so as to add greater rigor to empirical investigations in what is a complex and challenging area of investigation.

In the next three papers (by Lancor; Niebert and Gropengießer; and Fuchs, the perspective of conceptual metaphor is used to characterize student conceptions across a number of scientific topics, to design instructional representations and to identify the inherent narrative structure of a scientific domain and make a curricular recommendation to use narrative to prepare children for science in the early years. In the first of these papers, Lancor extends her earlier work investigating the range of analogies and metaphors for energy used in teaching and by students across a range of introductory college courses including biology, chemistry and physics. In this work, she has used the theory of conceptual metaphor and has identified six conceptual metaphors for energy. All of these are versions of the metaphorical construal of energy as a substance, but each highlights and obscures subtly different aspects of the concept (Lancor, 2014a). Lancor (2014b) has used this framework to analyze analogies for energy generated by students in physics, chemistry and biology courses. In her contribution to this special issue, she investigates how undergraduate students taking an interdisciplinary general science course make use of metaphors for energy when explaining the role of energy in relation to radiation, transportation, generating electricity, earthquakes and the big bang theory. When comparing the results from Lancor (2014a), she finds that the same framework of conceptual metaphors can be used in this interdisciplinary context as well. She reports the patterns of use of the six metaphors by students across topics and compares these patterns to findings from the disciplinary contexts. In light of the results of her study, Lancor argues that the framework of six conceptual metaphors for energy that she has developed offers a potential analytical lens which can be used as a way to reveal students' conceptual understanding, suggesting that it can be used by teachers as a formative assessment tool.

Also using Lakoff and Johnson's conceptual metaphor framework, Harald Gropengießer and colleagues at the University of Hannover have developed a range of learning activities building on the idea of *experientialism* (Gropengießer, 2007; Riemeier & Gropengießer, 2008). Drawing on the prior work of Gropengießer and colleagues, Niebert, Marsch, and Treagust (2012) analyzed a wide range of instructional analogies and metaphors used in science texts and science education research studies, and argued that the ones that are particularly effective are those that make use of students' embodied personal experiences as source domains. In their contribution to this special issue, Niebert and Gropengießer present an analysis of student interviews on conceptions of the greenhouse effect, carbon cycle, cell division and neurobiology. Using a conceptual metaphor perspective, they identify students' understanding in these domains which span macrocosmic (e.g. as in climate change) and microscopic (e.g. as in cell division) scales. By identifying image schemas that students apply that lead them to misconceptions in these domains, the authors uncover the 'learning demand' in each case. They then use this information to design effective instructional representations and evaluate their effect on student understanding through teaching experiments. Niebert and Gropengießer argue that understanding difficult concepts in science and, by implication, the design of instructional representations need to rely on knowledge acquired at the mesocosmic level of our everyday experiences.

Based on several years of teaching experiences in the field of thermodynamics and drawing on the literature on embodied cognition, Hans Fuchs has developed a novel approach to thermodynamics instruction (1987, 2010). He has argued that image-schemata (Johnson, 1987), especially *force-dynamic gestalts*, are used in making sense of thermodynamic processes (Fuchs, 2007). In his theoretical contribution here, Fuchs describes how image schemas are used to frame everyday situations and thermodynamic processes as narratives (Bruner, 1996). Fuchs analyses a story for children about cold gripping a village during winter, Sadi Carnot's account of heating water in a kettle, and modern continuum thermodynamics. He uncovers the image schemas that structure understanding of heat in these different contexts and shows how they combine to frame these diverse situations as narratives. He concludes by arguing that this suggests a role for a particular use of stories in early science instruction that can prepare children for the scientific thinking they will be expected to develop at more advanced stages.

In their commentary, David Treagust and Reinders Duit reflect on the papers by Lancor, Niebert and Gropengießer, and Fuchs. Treagust and Duit present their comments on these papers in relation to their prior, highly influential work on metaphors, analogies and multiple representations in conceptual change. In their commentary, they distinguish the three papers in terms of the extent to which they depart from what they refer to as 'classical' views on conceptual change, to which they have both contributed. They comment that the paper by Lancor seems to fit most clearly within the classical tradition, given the absence of an explicit reference to embodiment in her use of the conceptual metaphor framework. In their view, the novelty of Lancor's research is in extending accounts of student conceptions of energy beyond physics to other domains like chemistry and biology as well as interdisciplinary contexts. In their comment on Niebert and Gropengießer's paper, they find praiseworthy the theoretical synthesis of an embodied cognition perspective with prior work on multiple representations in science. They find particularly interesting the connections made to evolutionary epistemology, through the idea that human perceptual systems were designed to represent the world at the intermediate scale of the mescosm. They see great promise in this perspective as it is able to lead to successful instructional interventions in challenging domains of science. In Fuchs' theoretical proposal, they see a perspective on narrative in science and science learning that goes beyond prior work on narrative in science education and provides insights into the roles of models in science, science teaching and learning. They comment, however, that the argument used for developing this perspective of 'narrative framing' is complex and would need to be presented in a simpler and clearer language if it is to be communicated effectively to teachers. Overall, Treagust and Duit view this collection of papers as providing a lot of information about 'conceptual metaphor in action' with all papers showing how a conceptual metaphor perspective can make a contribution to effective instruction in science. They suggest, however, that greater precision and consistency is needed in the use of the construct of conceptual metaphor across authors.

The final paper in this issue (by Amin) is a review of the literature on conceptual metaphor in science education, exploring its contributions to the study of conceptual change. Amin (2009) had made a case for the relevance of a conceptual metaphor perspective for understanding conceptual change. However, the last seven or eight years have seen the emergence of research programs investigating different aspects of conceptual metaphor in science education as is reflected in this special issue. In his contribution to this issue, Amin reviews this literature with the specific goal of clarifying its contributions to the study of conceptual change. This paper first draws on Amin, Smith, and Wiser (2014) to present a highly condensed historical overview of research on conceptual change. The literature on conceptual metaphor is then reviewed so as to clarify its contributions to characterizing student misconceptions, identifying obstacles to learning, characterizing the process of conceptual change and designing effective instruction. However, Amin points out differences among researchers investigating conceptual metaphor in science education and suggests that this perspective has still not provided an explicit account of concepts. He briefly presents a way of viewing concepts while incorporating attention to the phenomenon of conceptual metaphor. He suggests directions for future research using a conceptual metaphor perspective that could contribute further to the study of conceptual change.

We hope that by bringing together these papers and commentaries on conceptual metaphor and embodied cognition in science learning in this issue, we will encourage further exploration, discussion and debate regarding the issues raised in its pages.

## Acknowledgements

This special issue grew out of a symposium on Conceptual Metaphor and Embodied Cognition in Science Learning at the European Science Education Research Association (ESERA) in September 2013. David Treagust graciously agreed to serve as discussant in this symposium. It was his formal discussant's reflections on the papers included in the symposium and our informal discussions with him afterwards that led to this project. We are grateful to David for encouraging us to put together this special issue.

## Disclosure statement

No potential conflict of interest was reported by the authors.

## References

Alibali, M. W., & Nathan, M. J. (2012). Embodiment in mathematics teaching and learning: Evidence from learners' and teachers' gestures. *Journal of the Learning Sciences, 21*(2), 247–286.

Amin, T. G. (2009). Conceptual metaphor meets conceptual change. *Human Development, 52*(3), 165–197.

Amin, T. G., Jeppsson, F., Haglund, J., & Strömdahl, H. (2012). The arrow of time: Metaphorical construals of entropy and the second law of thermodynamics. *Science Education, 96*(5), 818–848.

Amin, T. G., Smith, C. L., & Wiser, M. (2014). Student conceptions and conceptual change: Three overlapping phases of research. In N. G. Lederman & S. K. Abell (Eds.), *Handbook of research in science education* (Vol. 2, pp. 57–81). New York, NY: Routledge.

Andersson, B. (1986). The experiential gestalt of causation: A common core to pupils' preconceptions in science. *European Journal of Science Education, 8*(2), 155–171.

Barsalou, L. (2008). Grounded cognition. *Annual Review of Psychology*, 59(1), 617–645.

Brookes, D. T., & Etkina, E. (2007). Using conceptual metaphor and functional grammar to explore how language used in physics affects student learning. *Physical Review Special Topics—Physics Education Research*, 3(1), 010105.

Brookes, D. T., & Etkina, E. (2009). 'Force,' ontology, and language. *Physical Review Special Topics—Physics Education Research*, 5(1), 010110.

Bruner, J. (1996). *The culture of education*. Cambridge, MA: Harvard University Press.

Chi, M. T. H., Feltovich, P. J., & Glaser, R. (1981). Categorization and representation of physics problems by experts and novices. *Cognitive Science*, 5(2), 121–152.

Chi, M. T. H., & Slotta, J. D. (1993). The ontological coherence of intuitive physics. *Cognition and Instruction*, 10(2–3), 249–260.

Chi, M. T. H., Slotta, J. D., & De Leeuw, N. (1994). From things to processes: A theory of conceptual change for learning science concepts. *Learning and Instruction*, 4(1), 27–43.

Clark, A., & Chalmers, D. (1998). The extended mind. *Analysis*, 58(1), 7–19.

Clement, J. (2009). *Creative model construction in scientists and students: The role of imagery, analogy, and mental simulation*. Dordrecht: Springer.

Dewey, J. (1916). *Democracy and education: An introduction to the philosophy of education*. New York, NY: McMillan.

Dreyfus, B. W., Geller, B. D., Gouvea, J., Sawtelle, V., Turpen, C., & Redish, E. (2014). Ontological metaphors for negative energy in an interdisciplinary context. *Physical Review Special Topics—Physics Education Research*, 10(2), 020108.

Fauconnier, G., & Turner, M. (1998). Conceptual integration networks. *Cognitive Science*, 22(2), 133–187.

Fauconnier, G., & Turner, M. (2002). *The way we think: Conceptual blending and the mind's hidden complexities*. New York, NY: Basic Books.

Fuchs, H. U. (1987). Entropy in the teaching of introductory thermodynamics. *American Journal of Physics*, 55(3), 215–219.

Fuchs, H. U. (2007). *From image schemas to dynamical models in fluids, electricity, heat and motion. An essay on physics education research*. Retrieved October 26, 2012, from https://home.zhaw.ch/~fusa/COURSES/JO/Files_V/PER_Essay.pdf

Fuchs, H. U. (2010). *The dynamics of heat: A unified approach to thermodynamics and heat transfer* (2nd ed.). New York, NY: Springer.

Gallese, V., & Lakoff, G. (2005). The brain's concepts: The role of the sensory-motor system in conceptual knowledge. *Cognitive Neuropsychology*, 22(3–4), 455–479.

Gentner, D., Bowdle, B. F., Wolff, P., & Boronat, C. (2001). Metaphor is like analogy. In D. Gentner, K. J. Holyoak, & B. K. Kokinov (Eds.), *The analogical mind: Perspectives from cognitive science* (pp. 199–254). Cambridge, MA: MIT Press.

Gibbs, R. W. (2005). *Embodiment and cognitive science*. Cambridge: Cambridge University Press.

Gibson, J. J. (1979). *The ecological approach to visual perception*. Boston, MA: Houghton Mifflin.

Glenberg, A. M. (2008). Embodiment for education. In P. Calvo & T. Gomila (Eds.), *Handbook of cognitive science: An embodied approach* (pp. 355–372). Oxford: Elsevier Science.

Gropengießer, H. (2007). Theorie des erfahrungsbasierten Verstehens [The theory of experientialism]. In D. Krüger & H. Vogt (Eds.), *Theorien in der Biologiedidaktischen Forschung* (pp. 105–116). Berlin: Springer.

Gupta, A., Hammer, D., & Redish, E. F. (2010). The case for dynamic models of learners' ontologies in physics. *Journal of the Learning Sciences*, 19(3), 285–321.

Hall, R., & Nemirovsky, R. (2012). Introduction to the special issue: Modalities of body engagement in mathematical activity and learning. *Journal of the Learning Sciences*, 21(2), 207–215.

Hutchins, E. (1995). *Cognition in the wild*. Cambridge, MA: MIT Press.

Jeppsson, F., Haglund, J., Amin, T. G., & Strömdahl, H. (2013). Exploring the use of conceptual metaphors in solving problems on entropy. *Journal of the Learning Sciences*, 22(1), 70–120.

Johnson, M. (1987). *The body in the mind: The bodily basis of meaning, imagination, and reason.* Chicago, IL: University of Chicago Press.

Kontra, C., Goldin-Meadow, S., & Beilock, S. L. (2012). Embodied learning across the life span. *Topics in Cognitive Science, 4*(4), 731–739.

Lakoff, G., & Johnson, M. (1980). *Metaphors we live by.* Chicago, IL: University of Chicago Press.

Lakoff, G., & Johnson, M. (1999). *Philosophy in the flesh.* New York, NY: Basic Books.

Lakoff, G., & Núñez, R. E. (2000). *Where mathematics comes from: How the embodied mind brings mathematics into being.* New York, NY: Basic Books.

Lancor, R. A. (2014a). Using metaphor theory to examine conceptions of energy in biology, chemistry, and physics. *Science & Education, 23*(6), 1245–1267.

Lancor, R. A. (2014b). Using student-generated analogies to investigate conceptions of energy: A multidisciplinary study. *International Journal of Science Education, 36*(1), 1–23.

Lemke, J. L. (1998). *Teaching all the languages of science: Words, symbols, images and actions.* Retrieved 10 January, 2011, from http://academic.brooklyn.cuny.edu/education/jlemke/papers/barcelon.htm

Merleau-Ponty, M. (1962/2002). *Phenomenology of perception* (C. Smith, Trans.). London: Routledge.

Nemirovsky, R., Rasmussen, C., Sweeney, G., & Wawro, M. (2012). When the classroom floor becomes the complex plane: Addition and multiplication as ways of bodily navigation. *Journal of the Learning Sciences, 21*(2), 287–323.

Niebert, K., Marsch, S., & Treagust, D. F. (2012). Understanding needs embodiment: A theory-guided reanalysis of the role of metaphors and analogies in understanding science. *Science Education, 96*(5), 849–877.

Núñez, R. E., & Sweetser, E. (2006). With the future behind them: Convergent evidence from Aymara language and gesture in the crosslinguistic comparison of spatial construals of time. *Cognitive Science, 30*(3), 401–450.

Podolefsky, N. S., & Finkelstein, N. D. (2007). Analogical scaffolding and the learning of abstract ideas in physics: An example from electromagnetic waves. *Physical Review Special Topics— Physics Education Research, 3*(1), 010109.

Riemeier, T., & Gropengießer, H. (2008). On the roots of difficulties in learning about cell division: Process-based analysis of students' conceptual development in teaching experiments. *International Journal of Science Education, 30*(7), 923–939.

Scherr, R. E., Close, H. G., Close, E. W., Flood, V. J., McKagan, S. B., Robertson, A. D., ... Vokos, S. (2013). Negotiating energy dynamics through embodied action in a materially structured environment. *Physical Review Special Topics—Physics Education Research, 9*(2), 020105.

Scherr, R. E., Close, H. G., McKagan, S. B., & Vokos, S. (2012). Representing energy. I. Representing a substance ontology for energy. *Physical Review Special Topics— Physics Education Research, 8*(2), 020114.

Sfard, A. (1994). Reification as the birth of metaphor. *For the Learning of Mathematics, 14*(1), 44–55.

Shapiro, L. A. (2011). *Embodied cognition.* New York, NY: Routledge.

Turner, M., & Fauconnier, G. (1995). Conceptual integration and formal expression. *Metaphor and Symbolic Activity, 10*(3), 183–204.

Varela, F. J., Thompson, E., & Rosch, E. (1991). *The embodied mind: Cognitive science and human experience.* Cambridge, MA: MIT Press.

Vygotsky, L. S. (1978). *Mind in society: The development of higher psychological processes.* Cambridge, MA: Harvard University Press.

Williams, R. F. (2012). Image schemas in clock-reading: Latent errors and emerging expertise. *Journal of the Learning Sciences, 21*(2), 216–246.

Wilson, M. (2002). Six views of embodied cognition. *Psychonomic Bulletin & Review, 9*(4), 625–636.

Wittmann, M. C. (2010). *Using conceptual blending to describe emergent meaning in wave propagation.* Paper presented at the 9th International Conference of the Learning Sciences, ICLS '10, Chicago, IL. Retrieved from http://arxiv.org/abs/1008.0216

# The Importance of Language in Students' Reasoning About Heat in Thermodynamic Processes

David T. Brookes[a] and Eugenia Etkina[b]

[a]*Department of Physics, Florida International University, Miami, FL, USA;*
[b]*The Graduate School of Education, Rutgers University, New Brunswick, NJ, USA*

Researchers believe that the way that students talk, specifically the language that they use, can offer a window into their reasoning processes. Yet the connection between what students are saying and what they are actually thinking can be ambiguous. We present the results of an exploratory interview study with 10 participants, designed to investigate the role of language in university physics students' reasoning about heat in thermodynamic processes. The study revealed two key findings: (1) students' approaches to solving certain heat-related problems are related to the way in which they explicitly define the word 'heat' and (2) students' tendency to reason with heat as a state function in inappropriate contexts appears to be connected to a model of heat implicitly encoded in language. This model represents heat or heat energy/thermal energy as a substance that moves from one location to another. In this model, students talk about thermodynamic systems as 'containers' of heat, and temperature is a measure of the amount of heat 'in' an object.

In this paper we will explore the interplay between how university physics students speak and how they reason about heat in thermodynamics processes. We will focus on one particular area of student reasoning in thermodynamics: several studies in physics and chemistry education have found that there is a recurring pattern in student reasoning about thermodynamics processes. Specifically, a majority of university-level physics and chemistry students conceptualize heat as having the characteristics of a state function (an extensive thermodynamic quantity that is independent of thermodynamic path) (Fuchs, 1987; Kaper & Goedhart, 2002; Loverude, Kautz, & Heron, 2002; Meltzer, 2004; Roon, Sprang, & Verdonk, 1994). For example, in an interview study of 32 students enrolled in a calculus-based introductory university physics course, Meltzer (2004) found that 69% of the interviewees said that the total heat transfer for a closed thermodynamic cycle was zero. He observed that students most commonly argued (incorrectly) that the heat transferred into and out of the system during the cycle would be the same because the initial and final temperatures of the system were the same. Students focused on the beginning and end points of the process and ignored the path that was taken. This is the essence of state function-like reasoning.

Possibly related to this reasoning is the way in which we talk about heat in everyday language and even in well-established scientific fields such as physics and chemistry. It is common to talk about heat as a substance (Chi & Slotta, 1993; Chi, Slotta, & de Leeuw, 1994; Reiner, Slotta, Chi, & Resnick, 2000) that flows or is transferred into and out of thermodynamic systems. Thermodynamic systems are seen as containers of heat (Engel Clough & Driver, 1985; Erickson, 1979; Fuchs, 1987; Jasien & Oberem, 2002; Kesidou & Duit, 1993; Warren, 1972) and the temperature as an indicator of the amount of heat in the system (Beall, 1994; Erickson, 1979; Jasien & Oberem, 2002; Kesidou & Duit, 1993; Rozier & Viennot, 1991; Tripp, 1976; Warren, 1972). We will call this way of speaking, and its implicit ontology of heat as a substance and objects as containers of heat, the caloric metaphor. (The word caloric is a historical term from the eighteenth century, coined by Lavoisier, given to a hypothetical fluid that is transferred by contact between objects at different temperatures.) In prior research we found this caloric metaphor used roughly 40–50% of the time in all sentences that contained the word 'heat' in popular university-level introductory physics textbooks (Brookes, Horton, Van Heuvelen, & Etkina, 2005). In the view of Jeppsson, Haglund, Amin, and Strömdahl (2013), conceptual metaphors such as the caloric metaphor should be treated as resources that students can recruit in their reasoning. Brewe (2011) has suggested that thinking about kinetic or potential energy as a substance that flows and is stored in systems is a productive resource for students (note that, for example, gravitational potential energy *is* a state function). However, talking about heat this way is frequently blamed for students' difficulties with heat in thermodynamic processes (Arnold & Millar, 1994; Bauman, 1992; Brown, 1950; Fuchs, 1987; Harris, 1981; Heath, 1974; Hobson, 1995; Leff, 1995; Pushkin, 1996; Tripp, 1976; Zemansky, 1970).

It is likely difficult or impossible to conduct a controlled experimental study to see if there is an unambiguous causal link between language and reasoning. In other words, does the language we use incline reasoners to think about heat as a state function? The challenge stems from the difficulty of expunging or being able to counteract the caloric metaphor in a treatment group. The evidence of a causal connection is only suggestive: Hewson (1984) showed that first language Sotho speakers in South Africa, whose language possesses a dominant cultural metaphor in which 'heat' or the state of being hot is associated with emotional agitation or sickness, reasoned about heat in a way that was more closely aligned with kinetic molecular theory and rarely invoked caloric ideas. The authors explicitly shy away from concluding that language is influencing reasoning. Instead, they emphasize that the interaction between language and physical experience is a bi-directional one in which each can change and influence the other. In a more recent study, Kaper and Goedhart (2002) gave two groups of five students two different ways to talk about energy. The first group used the traditional 'forms of energy' language, while the second group was introduced to a new 'exchange value' language. When both groups of students were presented with and asked to interpret some thermodynamic processes, they found that the 'exchange value' group were able to move beyond state function reasoning about heat in those processes while the 'forms of energy' group remained trapped in inappropriate state function reasoning. The authors acknowledge that the scale of their study is too small to draw firm conclusions about the influence of language on students' reasoning.

We share the view, originally formulated by Sapir (1957) and Whorf (1956), that our experience of the physical world can influence our language, but that language can also influence our interpretation and understanding of physical experience. This bi-directional interaction of language and experience serves as a fundamental theoretical assumption in our work.

In this paper, we will adopt a viewpoint in which students are able to recruit various resources (Hammer, 2000) when engaged in reasoning about a physics problem. In the case of thermodynamics, we wish to examine whether state function reasoning is supported by a linguistic resource, namely the caloric metaphor. If this is true, we should see evidence that students who display state function-like reasoning recruit that metaphor more frequently than those who do not. Our prior research suggests additional complexity: physical models encoded in language have both explicit and implicit ontological components (Brookes & Etkina, 2007, 2009). In the case of heat, the explicit ontological component can be found in students' responses to the direct question: 'What is heat, or how do you define heat?' The implicit component may be uncovered by examining the grammatical structure of the metaphors that students use when they talk about heat while explaining their reasoning. These two components should combine together into a locally coherent model of heat encoded in language. In this manuscript we will describe an interview study using Meltzer's (2004) interview questions that set out to examine both the explicit and implicit ontological components of students' language about heat and how these relate to their difficulties with the concept of heat as a state function while solving thermodynamics problems. The study attempts to answer the following question. Is there a relationship between the way students talk and the way they approach physical situations and problems?

## Theoretical Background

### Heat and Temperature in Science

As the notions of heat and temperature are the foundational concepts of this paper, we provide a brief overview of expert understanding of these concepts in thermodynamic systems. Examining expert understanding will also give us key insights into the relationship between how experts define heat and the conceptual metaphors that they use to talk about it. There is no definition of what heat is that all physicists and chemists will agree on. However, there is certainly a consensus view about what heat is not. While the physical quantity of heat ($Q$) is measured in the SI unit of Joules and is therefore a quantity of energy, heat is not a form of energy in the same sense as we think of kinetic energy or potential energy. Additionally, temperature is not a measure of the amount of heat in a thermodynamic system.

A thermodynamic system may be described by either microscopic state variables (e.g. the positions and momenta of all the molecules in the system) or by macroscopic state variables such as volume, pressure, and temperature. State functions such as the internal energy, or molecular kinetic energy of the thermodynamic system, are functions of these state variables. Physically, this means that if one specifies the values of the state variables of a thermodynamic system, a state function will return a unique value associated with that system configuration. For example, given the temperature of a monatomic ideal gas, one can say that the average kinetic energy of a molecule in the gas will be $K = (3/2)kT$. Consequently the internal energy of that gas, which is the sum of the kinetic energies of the individual molecules, is a state function as well. Since there is no such function associated with heat, one cannot think of heat as a state function or as a form of energy in the conventional sense. It is meaningless to ask 'what is the heat of the system' for a specified volume, temperature, and/or pressure. Heat represents a quantity of energy added to or taken from a thermodynamic system by a heating process such as placing a sealed metal container of room temperature air on a stove and raising its temperature by heating it, or placing the same metal container of air in a refrigerator, thereby cooling it down.

The role of heat in thermodynamics is better understood if we examine the equation describing the first law of thermodynamics: $\Delta U_{int} = Q + W$. This equation describes the behavior of a state function, the internal energy of a thermodynamic system ($U_{int}$). The internal energy can change through the addition or removal of energy from the system through two distinct processes, either heating or doing work. Heating processes are those that involve transfers of energy through 'thermal contact between the system and the surrounding environment'. Work processes involve a collective macroscopic action applied to the system or performed by the system. For example, a work process could involve compressing a gas by squeezing it with a piston. The symbols $Q$ and $W$ represent the quantities of energy (most often called 'heat' and 'work', respectively) that are added to or removed from the system by these two processes. The equation that represents the first law of thermodynamics can be confusing because, while $U_{int}$, $Q$, and $W$ are all quantities of energy, $U_{int}$ is a state function of the thermodynamic system. In contrast, $Q$ and $W$ are process variables that quantify additions or subtractions of energy from the system, resulting in changes in the state function $U_{int}$.

## Language Used to Model the Physical World

Humans use language to describe and model their physical experiences (Halliday, 1985). This same modeling function is key to understanding how language works in science, both when scientists talk to each other and when teachers communicate scientific ideas to their students in a science classroom (Lemke, 1990, 2004). To understand the modeling function of language, we have adopted a number of views from different areas of linguistics and cognitive science. In this section we will introduce these different views and show how they fit together to describe language in a physics classroom.

In prior research we have adopted the framework of Lakoff and Johnson (1980) that views much of language and thought as metaphorical. Human language consists of many expressions that do not initially appear to be figurative but on careful analysis reflect underlying conceptual metaphors. For example, consult any English language manual on how to insulate your house and you will find sentences such as 'in hot weather, heat invades from the outdoors' or 'better insulation keeps more heat in during the cold weather'. Sentences such as these suggest that 'heat' is a substance (sometimes a fluid) that moves, flows, or is transferred from one location to another and that objects/locations (such as your house or the outdoors) are 'containers' of heat. Sutton (1993) explains the presence of conceptual metaphors in physics in terms of their analogical origins. For example, the caloric metaphor entered the language of physics after the adoption of the caloric model of heat. Because of the analogical origins of the caloric metaphor (and similar metaphorical language), experts are aware of the limitations and applicability of their language models.

In previous research we have shown that physicists use metaphorical language as a representation of a physical model of the world. For example, physicists use a metaphor of 'tunneling' to describe the process by which a bound quantum mechanical object (e.g. an electron) can 'pass through' a potential 'barrier' without having enough energy to 'pass over' it. While these metaphors may take on subtly different forms, sometimes 'tunneling', sometimes 'leaking', they have systematic grammatical patterns that encode an implicit ontology of *matter* (nouns or noun groups), *processes* (verbs or verb groups), and *physical states* (grammatical 'location': essentially container metaphors denoted by prepositional phrases that begin with prepositions such as 'in', and 'into' and may be supported by prior choice of verb group such as 'is absorbed') (Brookes, 2006; Brookes & Etkina, 2007; Chi & Slotta, 1993; Chi et al., 1994). For example, when writing about $\alpha$-particles, Feynman, Leighton, and Sands (1965, pp. 7–8) state that ' ... they start out with the energy $E$ inside the nucleus and "leak" through the potential barrier'. Here the objects or substances in the model are the $\alpha$-particles (implicitly referred to by 'they'), the nucleus, and the potential barrier; all functioning grammatically as nouns or noun groups. The process is one of 'leaking through' the barrier, and the physical state of the system is denoted by the metaphor of the $\alpha$-particle contained within the nucleus, identified by 'inside' and 'through'.

With the caloric metaphor, different choices of verb may associate different meanings with heat. 'Flows' may suggest heat is a fluid, while 'is transferred' leaves the exact nature of heat somewhat ambiguous. Yet, underpinning all these sentences is a common grammatical structure: the grammar of the caloric metaphor involves (a) heat functioning as a noun, (b) a verb that implies some sort of movement of heat, followed by (c) one or two grammatical location structures. See Table 1 for examples. This analysis allows us to systematically identify the implicit ontology of heat and its role in thermodynamic systems in students' discourse. A full discussion of the grammar on which this analysis is based is covered in Halliday (1985). The connection between grammar and the ontology of physical models is discussed in Brookes and Etkina (2007).

*Dynamic Ontologies and Local Coherence*

We have shown in previous research in quantum mechanics that expert physicists change their language model (and ontology) quickly and easily when confronted with anomalous situations (Brookes & Etkina, 2007). Other researchers have observed that physics students' reasoning seems to 'straddle' ontological categories (Gupta, Hammer, & Redish, 2010) and that students are able to use conceptual metaphors with a seemingly 'incorrect' implicit ontology to reason productively about physical situations (Gupta, Elby, & Conlin, 2014). These viewpoints contrast with the work of Chi and Slotta (1993) and Chi et al. (1994) who have argued the case that ontology is more static and that conceptual change in physics (moving from novice to expert) requires that the reasoner recategorize a physics concept into a different ontological category. For example, as they become more expert reasoners in physics, students need to undergo a conceptual shift, moving the concept of heat from the ontological category of substances to the ontological category of processes before they can reason effectively about heat in thermodynamic processes. Similar to Jeppsson et al. (2013) we suggest that there is a middle ground in this debate. Conceptual metaphors and their implicit ontology are resources (Jeppsson et al., 2013). While experts are able to shift easily between different conceptual metaphors, seemingly playing 'fast and loose' with ontology, there is at least some local coherence: for some period of time a coherent model with a consistent set of ontological categories is being activated. Novice physics students, on the other hand, have the challenge of navigating expert ways of speaking that must appear ontologically 'chaotic' at

Table 1.  Grammatical structure of the caloric metaphor

| Grammatical function | Participant (noun/noun group) | Process (verb/verb group) | Location | Location |
|---|---|---|---|---|
| Common words | Heat/heat energy/ thermal energy | Flows/moves/is transferred/is rejected/is absorbed | Into/to/from/ out of object A | Into/to/from/ out of object B |

first. It is a necessary part of learning physics that students be given time and space to explore how particular models (and their oft-conflicting ontologies) can be used productively in different situations (Brookes & Etkina, 2009).

The local coherence of experts' ontology is sometimes hard to tease apart. This is especially true in thermodynamics and so we will consider the case of 'heat' in detail. In the context of heat in thermodynamics, Slisko and Dykstra (1997) have observed, there is little agreement between experts about what 'heat' means and its role in thermodynamic processes. Many experts say that heat is a form of energy (Brown, 1950; Lewis & Linn, 1996); however most of these researchers and teachers qualify this statement by stating that heat is a 'special' form of energy. Heat is generally either referred to as 'disordered/random' energy (Helsdon, 1976) or as 'moving energy'/ 'energy in transit' (Tripp, 1976; Zemansky, 1970). Doige and Day (2012) call these definitions 'Class II' and 'Class I', respectively. The Class I definition, 'heat is energy in transit', is the definition that dominates most of the university introductory level physics textbooks (Cutnell & Johnson, 2001; Giancoli, 2000; Halliday, Resnick, & Walker, 2003; Tipler, 1999; Walker, 2002). Note that heat cannot be a 'normal' form of energy because it is not a state function. The quantity of heat $Q$ represents how much energy is added to or removed from a thermodynamic system in a heating process. Uniformly these researchers and textbook authors invoke the caloric metaphor, writing about heat as a noun, matching the way in which physicists talk about energy: i.e. 'heat is transferred from A to B' is the same as 'energy is transferred from A to B'. Note that this way of speaking hides the fact that heat is a 'special' form of energy.

While beyond the scope of this paper, we speculate that there are two possible components to this very common choice of language: (a) one historical and (b) one cognitive. (a) As Sutton (1993) has pointed out, historical models seem to live on in modern scientists' language as conceptual metaphors. As we have discussed in earlier work (Brookes & Etkina, 2007) conceptual metaphors can encode productive modes of reasoning. There are many cases when it is productive to think of heat as a state function. (b) There is an inevitable ontological tension between the *process* of heating and the *amount* of energy added by that heating process. Expert reasoners may tend to talk about heat as a substance because there is a practical need to be able to talk about the *physical quantity Q* in the equation of the first law of thermodynamics. This quantity is referred to as 'heat', representing the amount of energy that is transferred in a particular heating process.

Other researchers have, however, suggested that, to avoid confusion, 'heat' should be defined exclusively as a process, a means by which energy is transferred from one place to another (Baierlein, 1994; Bauman, 1992; Heath, 1974, 1976; Hobson, 1995; Pushkin, 1997; Romer, 2001). Without exception, these researchers and teachers have suggested that the only acceptable way to talk about heat grammatically is as a verb (the fire heated the room) or as grammatical manner ('energy flowed into the chamber by heat/heating'). The process definition and matching grammatical usage is found relatively rarely in introductory university physics textbooks (Etkina, Gentile, & Van Huevelen, 2014; Serway & Beichner, 2000).

It is important to note that surveying textbooks and the published recommendations of educators and practitioners likely does not reflect the real-world language usage of experts reasoning about heat in thermodynamic processes. The empirical results of Jeppsson et al. (2013) support a more dynamical view of expert language in action, in contrast with the more rigid view that expert reasoners offer when asked to think and reflect about their language usage in formal publications. Apart from this study, we do not know of any study that examines *in vivo* language usage of *experts* solving thermodynamics problems. This should be a topic for future research.

The goal of this section is to illustrate that, to understand how language models are applied in physical reasoning and the local coherence of their ontology, we need to examine both the explicit ontology of the concept as the reasoner defines it *and* the implicit ontology encoded in the grammatical structure of the conceptual metaphors that the reasoner uses to talk about the concept.

This discussion allows us to clarify the research question we wish to address in this paper: Is there a relationship between the way students talk and the way they approach physical situations and problems in thermodynamics? Specifically, we want to know (1) When students activate state function reasoning, do they talk about heat with an identifiable metaphor? (2) Are students who are able to access more sophisticated expert-like meanings associated with 'heat' reasoning differently about heat in thermodynamic processes as compared to those who can only access more novice-like meanings? In reviewing the literature, there appears to be a consistent pattern between how experts conceptualize heat on the one hand and how they recommend we define it and talk about it on the other. But how does that play out in the real world of student reasoning and problem-solving? As researchers, we would like to believe that listening to what students say can give us insights into their reasoning. If we can find evidence of a clear association between the way students talk and their reasoning, it would give us a deeper insight into the way in which we try to understand student reasoning through language. It would also give teachers fundamental tools to help them understand what their students are thinking as they try to help them diagnose and overcome their difficulties with various physics topics.

## Method

### Population

We recruited 10 students from a variety of physics courses at a large North Eastern University. They ranged from an introductory algebra-based sequence to honors calculus-based physics, to junior physics majors. The requirement was that students had already covered the thermodynamics section of the syllabus. All except S7 were native English speakers.

### Materials and Procedure

We used the same interview questions as in an interview study conducted by Meltzer (2004).[1] Meltzer's interview questions were particularly appropriate for our study

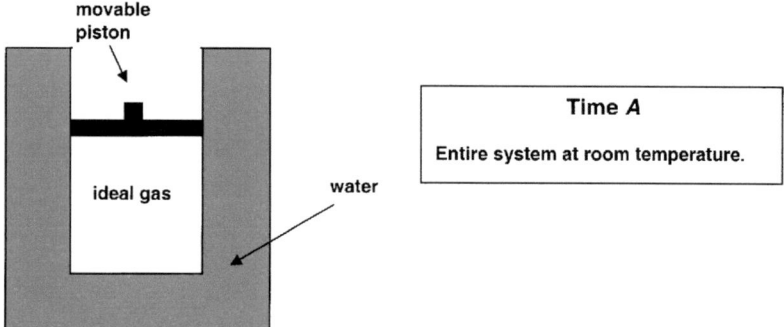

Figure 1.   The state of the system at the initial starting time A

because students had to answer questions about heat transfer, for both a closed thermodynamic cycle and in an isothermal step of that cycle. In addition we asked students to define the term 'heat' as best they could.

In the interview, each student was presented with a sheet of paper with a step-by-step description of a thermodynamic cycle performed by an ideal gas, enclosed in a piston and cylinder configuration. For each step of the cycle, students had to answer accompanying questions. The cylinder that enclosed the gas was depicted as having a jacket of water around it. This jacket served, in part, to control the rate of the thermodynamic processes so that the thermodynamic cycle could plausibly occur quasi-statically. Figure 1 shows the system at the starting point of the cycle.

In step 1 of the cycle from time A to time B, the piston was slowly raised by the gas at constant pressure as the gas was heated by the jacket that surrounded it. Figure 2 shows the state of the system after the initial heating step that raised the piston.

*Question 1*: During the process that occurs from time A to time B, which of the following is true: (a) positive work is done on the gas by the environment, (b) positive work is done by the gas on the environment, (c) no net work is done on or by the gas (correct answer is (b)).

*Question 2*: During the process that occurs from time A to time B, the gas absorbs $x$ J of energy from the water. Which of the following is true: the total kinetic energy of all of the gas molecules (a) increases by more than $x$ J; (b) increases by $x$ J; (c) increases,

Figure 2.   The state of the system at time B after the piston was raised by heating the gas

23

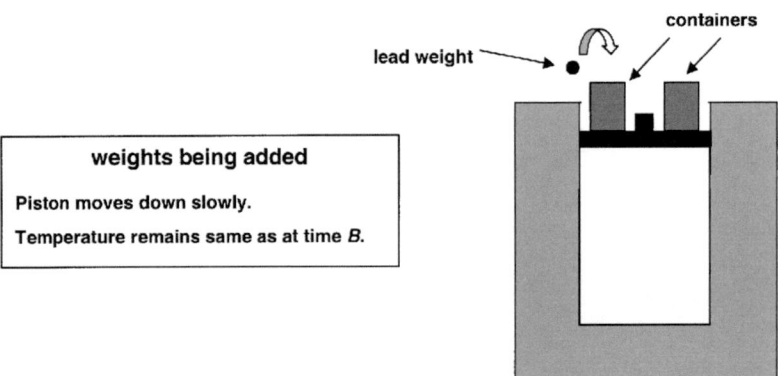

Figure 3.   Illustration of the containers placed on top of the piston as the B to C process begins

but by less than $x$ J; (d) remains unchanged; (e) decreases by less than $x$ J; (f) decreases by $x$ J; (g) decreases by more than $x$ J (correct answer is (c)).

In step 2, from time B to time C, the gas was slowly compressed by an external agent (containers were placed on top of the piston and small lead weights were gradually added to the containers). The act of adding lead weights is shown in Figure 3.

From time B to time C, the temperature was maintained at a constant value throughout this step of the process. Figure 4 shows the state of the system at time C.

*Question 3*: During the process that occurs from time B to time C, does the total kinetic energy of all the gas molecules increase, decrease, or remain unchanged? (Correct answer is 'remains unchanged').

*Question 4*: During the process that occurs from time B to time C, is there any net energy flow between the gas and the water? If no, explain why not. If yes, is there a net flow of energy from gas to water, or from water to gas?

The correct answer is that there should be energy flowing from the gas to the water ($Q$ gas → water) because as the gas is compressed, the total kinetic energy of the gas should increase. To keep the temperature constant, kinetic energy has to be removed from the gas as the compression takes place.

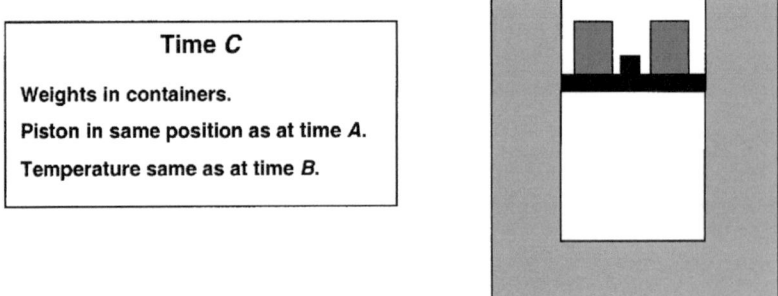

Figure 4.   The state of the system at time C

In the final step 3, students were told that the piston was locked in place and the gas was allowed to cool down to room temperature from time C to time D (a constant volume process). The final state of the system is shown in Figure 5. Then they were asked Q5 and Q6.

*Question 5*: During the process that occurs from time C to time D, the water absorbs $y$ J of energy from the gas. Which of the following is true: the total kinetic energy of all of the gas molecules (a) increases by more than $y$ J; (b) increases by $y$ J; (c) increases, but by less than $y$ J; (d) remains unchanged; (e) decreases, by less than $y$ J; (f) decreases by $y$ J; (g) decreases by more than $y$ J. Because the piston is locked in place there is no work done in this phase of the cycle. Consequently the energy transferred by heating must account for all of the energy that the gas loses. Thus the correct answer is (f).

*Question 6*: Consider the entire process from time A to time D. (i) Is the net work done by the gas on the environment during that process (a) greater than zero, (b) equal to zero, or (c) less than zero? (ii) Is the total heat transfer to the gas during that process (a) greater than zero, (b) equal to zero, or (c) less than zero? The correct answer to Q6(ii) is that the total heat transfer to the gas is less than zero ($Q < 0$). Physically this means that overall, for the gas to return to its initial pressure, temperature, and volume (a complete cycle), it had to get rid of some excess kinetic energy by heating.

After the thermodynamic cycle portion of the interview, we asked each student to discuss the meaning of the term 'heat'.

We recorded students with an mp3 recorder and an analog back-up recorder after one student's interview (S8) disappeared from the mp3 recorder before it could be saved. We only analyzed students' responses to Q4–Q6 since these questions (especially Q4 and Q6) required reasoners to go beyond state function reasoning with heat in order to answer correctly.

*Coding*

In our analysis, students' responses to Q4, Q5, and Q6(ii) were first coded for correctness and then examined for any patterns in their reasoning. We especially asked

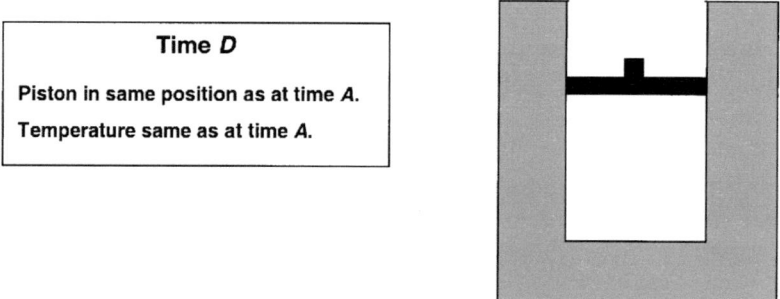

Figure 5.   The state of the system at time D

students to explain their reasoning in more detail when they came to Q6(ii). We rated students' level of understanding displayed in transcripts of their responses and justifications to Q6(ii) over and above their ability to get the right answer. This rating was done by a process of grouping responses with similar justifications and then ranking the different responses or groups of responses from weaker to better understanding of the problem. We also gave the student responses to a second rater who independently grouped students' responses and ranked students' understanding of the problem. There was 100% agreement between the two raters after discussion. A second analysis was then performed on Q6(ii) in which we counted the number of times the full caloric metaphor was invoked in each student's response and justification. For the full caloric metaphor to be identified we needed to identify (a) heat functioning as a noun, (b) a verb that described movement of heat, and (c) a grammatical location that heat was moving to or from. This coding was unequivocal, so we did not use a second coder for inter-rater reliability.

Finally, students' heat definitions were coded independently by two coders. Students' definitions were grouped into categories according to common features. Hundred percent agreement was achieved. Both coders participated in the creation of the coding scheme and thus were deeply familiar with it.


## Results and Analysis

Through the whole interview, we found a pattern of responses that was remarkably similar to those found by Meltzer (given the variability and small size of our population: algebra-based introductory physics students through to junior physics majors). In addition, many of the typical justifications were of a similar form (Meltzer, 2004).


### Students' Definitions of Heat

We were able to identify three different categories of heat definition. The first group consisted of those students who defined heat either as a substance on its own or as a form of energy without further elaboration. We classified their definitions together as a 'caloric/form of energy' definition. We grouped these definitions together because they possess a common trait: they do not identify subtleties of meaning that experts associate with the term 'heat'. (Physicists and chemists either treat heat as a 'special' form of energy, or as a 'process', a means of transferring energy from one location to another.) The second group consisted of those students who defined heat as 'energy in transit' or 'the quantity in the equation', similar to the operational definition given by Serway and Beichner (2000). We classified their definition as 'operational'. As mentioned above, this category of definition is similar to definitions put forth in most modern introductory college level physics textbooks. The third group consisted of those students who said that 'heat is a transfer of energy', rather than 'heat is energy that is transferred'. We classified their definitions as a 'process' definition. The three categories of definition with examples from the interviews are provided in Table 2.

Table 2.   All students' heat definitions from interview study

| Category | Student's definition |
|---|---|
| *Caloric* definition: Heat is a substance, or heat is a form of energy | S1: 'Temperature … is a measure of the heat … of the system'<br>S2: 'Heat is actual energy that gives the molecules the kinetic energy<br>S5: 'They [heat and temperature] are directly proportional. If you add heat, you increase temperature …'<br>S6: 'Heat is … the average kinetic energy, so it would be the total kinetic energy of the gas<br>Interviewer: 'Are you saying that the temperature is an indicator of the amount of heat in the system?' S9: 'Yeah, of course!' |
| *Operational* definition | S4: 'I'm not exactly sure what it means … for heat, all I know there is a specific quantity in the equation, I don't really understand what it is …'<br>S3: '[Heat is] just the energy that's transferred …' |
| *Process* definition | S7: '[Heat is] … not even a type of energy. A way of transferring energy from one system to another by thermal contact<br>S10: ' … it's a transfer, heat is a process, transferring energy …'<br>S8: [Process definition; reconstructed from interview notes] |

Table 3 shows all students' responses to Q4, Q5 and Q6(ii) of the interview study. These responses are tabulated with students' heat definitions.

In Q4, of the five students who defined heat as either a caloric substance or a form of energy, four said incorrectly that heat transfer was zero, while one student answered Q4 correctly. A typical justification for $Q = 0$ comes from S4:

> S4: There is no net flow of energy from the gas to the water because, um, the temperature remained the same and I guess that means there's no heat transferred.

Some of the other responses in this category included more convoluted reasoning, invoking the ideal gas law, but in each case, students used the ideal gas law to help them reason about the temperature of the system from B to C, rather than think-ing about the two competing processes (heating and doing work) that would cause the average kinetic energy (and consequently the temperature) of the gas to change.

Of the five students who defined heat as a 'special' form of energy (operational definition) or a process of energy transfer, one student said $Q = 0$ while four did

Table 3.  Summary of students' heat definitions and their responses to Q4, Q5, and Q6(ii) of the interview study

|     | Heat definition | Q4 ($Q$ gas $\rightarrow$ water) | Q5 ($y$ J) | Q6(ii) ($Q < 0$) |
| --- | --- | --- | --- | --- |
| S1  | Caloric/energy | $Q = 0$ | $y$ J | $Q = 0$ |
| S2  | Caloric/energy | $Q = 0$ | $y$ J | $Q = 0$ |
| S3  | Operational | $Q$ water $\rightarrow$ gas | $y$ J | $Q > 0$ |
| S4  | Operational | $Q = 0$ | $y$ J | Unsure |
| S5  | Caloric/energy | $Q = 0$ | $y$ J | $Q = 0$ |
| S6  | Caloric/energy | $Q = 0$ | $y$ J | $Q < 0$ |
| S7  | Process | $Q$ gas $\rightarrow$ water | $y$ J | $Q < 0$ |
| S8  | Process | $Q$ gas $\rightarrow$ water | $y$ J | $Q < 0$ |
| S9  | Caloric/energy | $Q$ gas $\rightarrow$ water | $y$ J | $Q = 0$ |
| S10 | Process | $Q$ gas $\rightarrow$ water | $y$ J | Unsure |

Note: Correct responses are shown in parentheses in the column headings.

not: $p = .103$, one-tailed Fisher exact test. An example of the most clear and succinct reasoning from this group comes from S7:

> S7: And in this process the internal energy doesn't change for the particles and there is some work performed to the system, so the system has to release energy through heating to the water.

In Q6(ii), of the five students who defined heat as either a substance or a form of energy, four said incorrectly that heat transfer was zero, while one student did not answer $Q = 0$. Of the five students who defined heat as a 'special' form of energy (operational definition) or a process of energy transfer, none of the students answered that $Q = 0$ for the complete thermodynamic cycle: $p = .024$, one-tailed Fisher exact test.

Interestingly, this association between heat definition and reasoning completely disappeared in Q5 where heating is the sole energy transfer process. However, we found an additional pattern of reasoning: students who could define heat as 'energy in transit or as a 'process' produced qualitatively different reasoning when justifying their answer to Q5. Students S3, S4, S7, and S10 explicitly mentioned that there was no work being done when answering Q5. A typical response was:

> S7: Well, the gas heats the water by y Joules, but there is no work done so the kinetic energy decreases by y Joules.

In contrast, none of the students who defined heat as a substance or a form of energy explicitly mentioned this fact (no work done). A typical response from this group was:

> S2:           ... okay. Decreases by y Joules.
> Interviewer:  Why do you say that?
> S2:           Um, because the net energy flow is to the water since the heat flow is to there as well and that's the only energy that's being transferred, so it's equivocal [sic].

This association between heat definition and justification on Q5 is also significant ($p = .024$, one-tailed Fisher exact test). Thus we see some evidence of distinct ideas being recruited when we examine the students' language.

## Student Responses to Q6(ii) and the Caloric Metaphor

As mentioned in section 'Method', we both ranked students' responses to Q6(ii) from poorer to better understanding, and counted the total number of occurrences of the caloric metaphor in their justification. The results of this coding are shown in Table 4.

Table 4 reveals a significant correlation between the quality of students' reasoning on Q6(ii) and the number of times they invoked the caloric metaphor in their explanation: $R_s = 0.85$, $p < .05$ (non-directional test).

Two examples of student responses are shown below. These are not the full responses (since they are too long). Students' responses to this question ranged from 780 words (S5) to 150 words (S10). In the following excerpt from S1, categorized as reflecting 'poor understanding' of Q6(ii), note the frequent use of the caloric metaphor (shown in italics):

> S1: And for part (ii) ... If it *returns back* to the same temperature, I would have to say, once again, its equal to zero. Well, the temperature increased, and then it decreased so now, so the net heat, like the *net heat transfer* I guess would be zero, and I guess, because it *goes up* then it *goes down back* to the same thing ...

Table 4. Students' recruitment of the caloric metaphor compared with their ability to understand Q6(ii)

| | Caloric metaphor count | Rank | Categories of student responses to Q6(ii) | Rank | Heat definition |
|---|---|---|---|---|---|
| S5 | 21 | 9 | Poor understanding ($Q = 0$) | 7.5 | Caloric/energy |
| S9 | 17 | 8 | Poor understanding ($Q = 0$) | 7.5 | Caloric/energy |
| S2 | 8 | 7 | Poor understanding ($Q = 0$) | 7.5 | Caloric/energy |
| S1 | 6 | 6 | Poor understanding ($Q = 0$) | 7.5 | Caloric/energy |
| S6 | 4 | 5 | Some idea | 5 | Caloric/energy |
| S4 | 3 | 3.5 | Good ideas | 3.5 | Operational |
| S10 | 0 | 1 | Good ideas | 3.5 | Process |
| S3 | 1 | 2 | Good understanding, just got signs mixed up | 2 | Operational |
| S7 | 3 | 3.5 | Best (correct, clear understanding) | 1 | Process |
| S8 | N/A | N/A | N/A | N/A | Process |

Note: These are shown with ranks used for Spearman's rank order correlation coefficient.

[Later] ... And that's what I would be thinking. Like because I think *it [heat] goes out as much as in* because if it returns back to the same temperature as it was at A, if *[heat] goes in* ... let's say it [the system] went up 10 degrees Kelvin, if it was up 10 degrees Kelvin and that's how the system changed. And then it loses that 10 degrees Kelvin, it's as though nothing ever happened. So I would say the total heat transfer is zero.

The following excerpt from S3 shows reasoning that was categorized as 'good understanding' of Q6(ii).

| S3: | [The total energy transfer] is also zero because it returns to the same temperature as it was at time A. If it returns to the same temperature, any energy it absorbed went back to the ... actually some of it went into work, but that also ... because the net work was zero ... So, if there was some energy that got transferred to the gas that was used in work and therefore ... there was positive energy flow to the gas. Does that make sense? ... some of the energy was used up so all the energy that was transferred into the gas by the water cannot possibly go back to the water. |
|---|---|
| Interviewer: | Okay. Where was it used up? |
| S3: | In moving the piston. |
| Interviewer: | So what would your answer be for the heat transfer. |
| S3: | I wanna say zero because it's the same temperature [laughs] [After a long pause] The *total heat transferred to the gas* is greater than zero. Because the gas does this work in that time and it needs energy to do that because it returns to the same temperature and position as it was at the beginning. In order to do that work it needs energy from the water. |

## Discussion

In this paper we set out to answer two research questions: (1) When students activate state function reasoning, do they talk about heat with an identifiable metaphor? (2) Are students who are able to access more sophisticated expert-like meanings associated with 'heat' reasoning differently about heat in thermodynamic processes as compared to those who can only access more novice-like meanings? While our sample size is small, we can see clear relationships between metaphors that students use, their understanding of the term 'heat', and their reasoning. (1) Students who incorrectly reasoned that $Q = 0$ in Q6(ii) invoked the caloric metaphor far more frequently in their justification than students who do not suggest $Q = 0$. (2) Students who defined heat as substance or a 'normal' form of energy were more likely to answer $Q = 0$ for Q4 and Q6(ii) as compared to students who were able to access more expert-like definitions of heat. On Q5 all students were able to get this question correct, independently of the model they were able to access. This is not so surprising since no work is done during this stage of the cycle. In this step of the thermodynamic cycle, thinking of heat as a state function would be a productive resource. More remarkably, students who could access an expert-like definition of heat all spontaneously mentioned that no work was done in this step. This suggests that their expert-like understanding of heat is accompanied by an overall more sophisticated understanding of thermodynamic processes, being able to recognize that there are two competing processes (work and heating) that add or subtract energy from the system.

With our data and methodology, it is not possible for us to make causal assertions about the impact of the caloric metaphor on students' reasoning, whether reasoning is being driven by language choices or if students' underlying conceptions of heat are influencing their choice of language. What we observe from our data is that the caloric metaphor is connected to reasoning about heat that appears as state function-like reasoning. For example, students who are applying the caloric metaphor are saying that if the thermodynamic system returns to its initial starting point (same temperature as before), the amount of heat/heat energy in the system should be the same as it was initially and thus the net heat transfer for a closed thermodynamic cycle should be zero. Likewise, if temperature stays constant (as in an isothermal process) there should be no heat transfer during the process.

There are a number of possible interpretations of these results that we would like to consider:

(1) It could appear to the reader that learning and the development of expertise is a progression from a substance-based conception of heat to a process conception of heat (Chi et al., 1994). However, we believe a more nuanced view is necessary. It is likely that the more expert reasoners who explicitly defined heat as a process still find it productive to access other conceptions of heat as a substance. Gupta et al. (2010) provide some empirical evidence in support of this interpretation. We suggest that experts simply have more linguistic resources to draw from than novices do.

(2) Controlled studies introducing students to alternative ways of talking about thermodynamic phenomena show that we can have an impact on students' understanding and reasoning in that domain (Kaper & Goedhart, 2002). This result supports the idea that instructional linguistic choices can influence how a physical phenomenon is conceptualized. Alternatively, studies show that students activate different linguistic resources depending on the phenomenon they are trying to explain (Gupta et al., 2010, 2014). This result suggests to us that particular thought processes are influencing the language that students use to describe the phenomenon. The fact that both of these perspectives are supported by empirical evidence suggests to us that, as we claimed earlier in the introduction, the connection between language and thought is a bi-directional one. This in turn leads directly to our third and final point.

(3) Lemke (2004) has suggested that understanding is simply the ability to coordinate multiple representations of a phenomenon (including language) in productive and effective ways. He suggests that concepts do not exist independently from their representations, and it is the representations themselves (graphs, equations, spoken/written language, etc.) that constitute the concept itself.

In the context of point 3 above, what we see as the significance of our research is that it deepens our understanding of the role of language as a semiotic resource in students' reasoning. We have shown how (a) implicit ontological metaphors (encoded in grammatical structures) and (b) explicit ontology as expressed in how physical terms such as 'heat' are defined by the reasoner play a role in the case of reasoning about heat in a thermodynamic process.

## Instructional Implications

Many physics textbooks explicitly tell students that heat is not like other forms of energy (rather, it is energy in transit) and is not a state function. And yet, any differentiation between heat as a regular form of energy and a 'special' form of energy ('disordered energy'/'energy in transit'), and/or heat as a process of energy transfer, is very subtle. It is therefore plausible that students entirely miss the subtle nuances of the experts' definition of heat (Roon et al., 1994) and simply stick to the idea that heat is a form of energy. This is especially complicated by the fact that most textbooks continue on talking about heat being transferred into or out of a thermodynamic system, entirely consistent with the 'energy in transit' definition. In this type of language, heat functions grammatically identically to any other form of energy. (In other words, the distinction between normal and special forms of energy cannot be made at a grammatical level.) We believe that the ubiquity of the caloric metaphor only encourages students to think of heat as a form of energy and suppress the nuances that are doubtless very difficult to understand on a single viewing of the definition.

While not directly related to the empirical findings in our paper, we would like to conclude by examining the tension that the bi-directional interaction of language and thought introduces into classroom discourse, and discuss a possible way to resolve that tension. To highlight this tension we observe that some researchers recommend that instructors be more precise in their language usage in the classroom (Williams, 1999), while others prefer to point out the value of students' innovative language usage even when that language is imprecise (Gupta et al., 2014). How can an instructor get students be more precise in their language usage without suppressing a classroom culture that values authentic sense-making? If we continue to follow the idea that available semiotic resources are what constitute a concept, learning physics is fundamentally an act of meaning-making (Lemke, 2004). From the observation of physical phenomena, we the instructors alongside with students develop semiotic resources (including language) to describe, make sense of, and then explain these observed phenomena. What that translates to in the classrooms of the authors is that we spend a great deal of effort as instructors, trying to create a classroom culture that adheres to certain linguistic norms. Specifically, students are not allowed to introduce technical terms before they establish an agreed meaning in the classroom learning community. The default expectation is that phenomena are described and explained in strictly non-technical terms. Introducing technical terms only happens later when the underlying mechanisms, the how and why of the phenomenon, is familiar to the members of the learning community. This approach is similar to what Arons (1997) has suggested, namely that technical nominalizing (e.g. calling a complex thermodynamic process 'heating') is one of the last stages of the learning process. Allowing language to naturally develop in this way, in our experience, helps students avoid many of the pitfalls that may arise from (a) linguistic resources that students bring into the classroom and whose meanings are not well established and (b) trying to explicitly define technical terms and encourage students to use the 'right' language from the get-go.

## Disclosure Statement

No potential conflict of interest was reported by the authors.

## Note

1. The interview questions and accompanying figures are reproduced with permission from Meltzer (2004), © 2004, American Association of Physics Teachers.

## References

Arnold, M., & Millar, R. (1994). Children's and lay adults' views about thermal equilibrium. *International Journal of Science Education, 16*(4), 405–419.

Arons, A. B. (1997). *Teaching introductory physics.* New York, NY: John Wiley and Sons, Inc.

Baierlein, R. (1994). Entropy and the second law: A pedagogical alternative. *American Journal of Physics, 62*(1), 15–26.

Bauman, R. P. (1992). Physics that textbook writers usually get wrong: II. Heat and energy. *The Physics Teacher, 30*(6), 353–356.

Beall, H. (1994). Probing student misconceptions in thermodynamics with in-class writing. *Journal of Chemical Education, 71*(12), 1056–1057.

Brewe, E. (2011). Energy as a substance like quantity that flows: Theoretical considerations and pedagogical consequences. *Physical Review Special Topics—Physics Education Research, 7*(2), 020106.

Brookes, D. T. (2006). *The role of language in learning physics* (Unpublished doctoral dissertation, Rutgers). The State University of New Jersey, New Brunswick.

Brookes, D. T., & Etkina, E. (2007). Using conceptual metaphor and functional grammar to explore how language used in physics affects student learning. *Physical Review Special Topics—Physics Education Research, 3*(1), 010105.

Brookes, D. T., & Etkina, E. (2009). "Force," ontology, and language. *Physical Review Special Topics—Physics Education Research, 5*(1), 010110.

Brookes, D. T., Horton, G. K., Van Heuvelen, A., & Etkina, E. (2005). Concerning scientific discourse about heat. In J. Marx, P. Heron, & S. Franklin (Eds.), *Proceedings of the 2004 physics education research conference* (Vol. 790, pp. 149–152). Melville, NY: American Institute of Physics.

Brown, S. C. (1950). The caloric theory of heat. *American Journal of Physics, 18*(6), 367–373.

Chi, M. T. H., & Slotta, J. D. (1993). The ontological coherence of intuitive physics. *Cognition and Instruction, 10*(2 & 3), 249–260.

Chi, M. T. H., Slotta, J. D., & de Leeuw, N. (1994). From things to processes: A theory of conceptual change for learning science concepts. *Learning and Instruction, 4*(1), 27–43.

Cutnell, J. D., & Johnson, K. W. (2001). *Physics* (5th ed.). New York, NY: John Wiley & Sons.

Doige, C. A., & Day, T. (2012). A typology of undergraduate textbook definitions of 'heat' across science disciplines. *International Journal of Science Education, 34*(5), 677–700.

Engel Clough, E., & Driver, R. (1985). Secondary students' conceptions of the conduction of heat: Bringing together scientific and personal views. *Physics Education, 20*(4), 176–182.

Erickson, G. L. (1979). Children's conceptions of heat and temperature. *Science Education, 63*(2), 221–230.

Etkina, E., Gentile, M., & Van Huevelen, A. (2014). *College physics*. Boston, MA: Pearson.

Feynman, R. P., Leighton, R. B., & Sands, M. (1965). *The Feynman lectures on physics: Quantum mechanics*. Reading, MA: Addison-Wesley.

Fuchs, H. U. (1987). Thermodynamics: A "misconceived" theory. In J. D. Novak (Ed.), *Proceedings of the second international seminar on misconceptions and educational strategies in science and mathematics* (Vol. 3, pp. 160–167). Ithaca, NY: Cornell University.

Giancoli, D. C. (2000). *Physics for scientists & engineers* (3rd ed.). Upper Saddle River, NJ: Prentice Hall.

Gupta, A., Elby, A., & Conlin, L. D. (2014). How substance-based ontologies for gravity can be productive: A case study. *Physical Review Special Topics—Physics Education Research, 10*(1), 010113.

Gupta, A., Hammer, D., & Redish, E. F. (2010). The case for dynamic models of learners' ontologies in physics. *Journal of the Learning Sciences, 19*(3), 285–321.

Halliday, D., Resnick, R., & Walker, J. (2003). *Fundamentals of physics* (6th ed.). New York, NY: John Wiley & Sons.

Halliday, M. A. K. (1985). *An introduction to functional grammar*. London: Edward Arnold.

Hammer, D. (2000). Student resources for learning introductory physics. *American Journal of Physics, 68*(S1), S52–S59.

Harris, W. F. (1981). Heat in undergraduate education, or isn't it time we abandoned the theory of caloric? *International Journal of Mechanical Engineering Education, 9*(4), 317–325.

Heath, N. E. (1974). Heating. *Physics Education, 9*(7), 490–491.

Heath, N. E. (1976). Heating. *Physics Education, 11*(6), 389–390.

Helsdon, R. M. (1976). Teaching thermodynamics. *Physics Education, 11*(4), 261–262.

Hewson, M. G. (1984). The influence of intellectual environment on conceptions of heat. *European Journal of Science Education, 6*(3), 245–262.

Hobson, A. (1995). Heat is not a noun. *The Physics Teacher, 33*(9), 325–326.

Jasien, P. G., & Oberem, G. E. (2002). Understanding of elementary concepts in heat and temperature among college students and K-12 teachers. *Journal of Chemical Education, 79*(7), 889–895.

Jeppsson, F., Haglund, J., Amin, T. G., & Strömdahl, H. (2013). Exploring the use of conceptual metaphors in solving problems on entropy. *Journal of the Learning Sciences, 22*(1), 70–120.

Kaper, W. H., & Goedhart, M. J. (2002). 'Forms of energy', an intermediary language on the road to thermodynamics? Part II. *International Journal of Science Education, 24*(2), 119–137.

Kesidou, S., & Duit, R. (1993). Students' conceptions of the second law of thermodynamics—An interpretive study. *Journal of Research in Science Teaching, 30*(1), 85–106.

Lakoff, G., & Johnson, M. (1980). *Metaphors we live by*. Chicago, IL: The University of Chicago Press.

Leff, H. S. (1995). Entropy and heat along reversible paths for fluids and magnets. *American Journal of Physics, 63*(9), 814–817.

Lemke, J. L. (1990). *Talking science: Language, learning, and values*. Norwood, NJ: Ablex Publishing Corporation.

Lemke, J. L. (2004). *Teaching all the languages of science: Words, symbols, images, and actions*. Presentation given at a conference, Barcelona. Retrieved from http://academic.brooklyn.cuny.edu/education/jlemke/sci-ed.htm

Lewis, E. L., & Linn, M. C. (1996). Where is the heat?: A response to David Pushkin. *Journal of Research in Science Teaching, 33*(3), 335–337.

Loverude, M. E., Kautz, C. H., & Heron, P. R. L. (2002). Student understanding of the first law of thermodynamics: Relating work to the adiabatic compression of an ideal gas. *American Journal of Physics, 70*(2), 137–148.

Meltzer, D. E. (2004). Investigation of students reasoning regarding heat, work, and the first law of thermodynamics in an introductory calculus-based general physics course. *American Journal of Physics, 72*(11), 1432–1446.

Pushkin, D. B. (1996). A comment on the need to use scientific terminology appropriately in conception studies. *Journal of Research in Science Teaching, 33*(2), 223–224.

Pushkin, D. B. (1997). Scientific terminology and context: How broad or narrow are our meanings? *Journal of Research in Science Teaching, 34*(6), 661–668.

Reiner, M., Slotta, J. D., Chi, M. T. H., & Resnick, L. B. (2000). Naive physics reasoning: A commitment to substance-based conceptions. *Cognition and Instruction, 18*(1), 1–34.

Romer, R. H. (2001). Heat is not a noun. *American Journal of Physics, 69*(2), 107–109.

van Roon, P. H., van Sprang, H. F., & Verdonk, A. H. (1994). 'Work' and 'heat': On a road towards thermodynamics. *International Journal of Science Education, 16*(2), 131–144.

Rozier, S., & Viennot, L. (1991). Students' reasonings in thermodynamics. *International Journal of Science Education, 13*(2), 159–170.

Sapir, E. (1957). The status of linguistics as a science. In D. B. Mandelbaum (Ed.), *Culture, language and personality* (pp. 65–77). Berkeley, CA: University of California Press.

Serway, R. A., & Beichner, R. J. (2000). *Physics for scientists and engineers* (5th ed., Vol. 1). Orlando, FL: Saunders College Publishing.

Slisko, J., & Dykstra, D. I. (1997). The role of scientific terminology in research and teaching: Is something important missing? *Journal of Research in Science Teaching, 34*(6), 655–660.

Sutton, C. R. (1993). Figuring out a scientific understanding. *Journal of Research in Science Teaching, 30*(10), 1215–1227.

Tipler, P. A. (1999). *Physics for scientists and engineers* (4th ed., Vol. 1). New York, NY: W. H. Freeman and Company/Worth Publishers.

Tripp, T. B. (1976). The definition of heat. *Journal of Chemical Education, 53*(12), 782–784.

Walker, J. S. (2002). *Physics.* Upper Saddle River, NJ: Prentice Hall.

Warren, J. W. (1972). The teaching of the concept of heat. *Physics Education, 7*(1), 41–44.

Whorf, B. L. (1956). The relation of habitual thought and behavior to language. In J. B. Carroll (Ed.), *Language, thought, and reality. Selected writings of Benjamin Lee Whorf* (pp. 134–159). Cambridge, MA: The MIT Press.

Williams, H. T. (1999). Semantics in teaching introductory physics. *American Journal of Physics, 67*(8), 670–680.

Zemansky, M. W. (1970). The use and misuse of the word "heat" in physics teaching. *The Physics Teacher, 8*(6), 295–300.

# Varying Use of Conceptual Metaphors across Levels of Expertise in Thermodynamics[†]

Fredrik Jeppsson[a], Jesper Haglund[b] and Tamer G. Amin[c]

[a]*Department of Social and Welfare Studies, Linköping University, Norrköping, Sweden;* [b]*Department of Physics and Astronomy, Uppsala University, Uppsala, Sweden;* [c]*Science and Mathematics Education Center, Department of Education, American University of Beirut, Beirut, Lebanon*

Many studies have previously focused on how people with different levels of expertise solve physics problems. In early work, focus was on characterising differences between experts and novices and a key finding was the central role that propositionally expressed principles and laws play in expert, but not novice, problem-solving. A more recent line of research has focused on characterising continuity between experts and novices at the level of non-propositional knowledge structures and processes such as image-schemas, imagistic simulation and analogical reasoning. This study contributes to an emerging literature addressing the coordination of both propositional and non-propositional knowledge structures and processes in the development of expertise. Specifically, in this paper, we compare problem-solving across two levels of expertise—undergraduate students of chemistry and Ph.D. students in physical chemistry—identifying differences in how conceptual metaphors (CMs) are used (or not) to coordinate propositional and non-propositional knowledge structures in the context of solving problems on entropy. It is hypothesised that the acquisition of expertise involves learning to coordinate the use of CMs to interpret propositional (linguistic and mathematical) knowledge and apply it to specific problem situations. Moreover, we suggest that with increasing expertise, the use of CMs involves a greater degree of subjective engagement with physical entities and processes. Implications for research on learning and instructional practice are discussed.

---

[†]Third contribution to special issue entitled: Conceptual metaphor and embodied cognition in science learning

## Introduction

The study of expertise in science has been a topic of investigation for some time. Various perspectives can be found in the literature. Early work on expertise focused on the differences between novices and experts, identifying the important role of abstract principles in the latter, absent in the former (Chi, 2006a, 2006b; Chi, Feltovich, & Glaser, 1981; Chi, Glaser, & Rees, 1982). Another line of work has emphasised the continuity between novices and experts by focusing on non-propositional knowledge structures such as imagery, image-schemas (abstractions from sensorimotor experiences) and mental models (Clement, 2009; diSessa, 1993; Hammer, 2000; Smith, diSessa, & Roschelle, 1993). This latter approach can be seen as reflecting a kind of 'embodiment turn' in research on scientific expertise and science learning. More recently, it has been increasingly acknowledged that scientific expertise involves the coordination of abstract knowledge, formulated propositionally, and non-propositional knowledge structures and processes (e.g. Amin, 2009; Amin, Smith, & Wiser, 2014; Cheng & Brown, 2010; Jeppsson, Haglund, Amin, & Strömdahl, 2013; Sherin, 2001, 2006). The study reported in this paper contributes to this recent effort aimed at understanding the difference between novices and experts in terms of both propositionally represented knowledge and non-formal knowledge structures and processes. We use the term propositional to refer to language-like representations (such as natural language and mathematical formalism) which bear no resemblance to what they represent and express a belief that can be judged true or false. We use the term non-propositional to refer to analogical representations that preserve structural features of what they represent, such as images, image-schematic abstractions from sensorimotor experience and mental models. In a previous study (Jeppsson et al., 2013), we investigated the role that systematic metaphorical mappings (conceptual metaphors, CMs), implicit in language, play in advanced problem-solving in thermodynamics. One of the findings in that study was that the advanced scientific problem-solving of Ph.D. students in physical chemistry involved the coordination of multiple image schemas with each other and with propositional knowledge expressed through language and mathematics. The present study extends this previous research by examining how this coordination differs in Ph.D. students and undergraduates. The purpose is to identify whether the acquisition of expertise in scientific problem-solving involves changes in how non-propositional and propositional knowledge structures are coordinated, with a particular emphasis on the changing roles of CMs.

In the rest of this introduction, we briefly review the literature on scientific expertise, reviewing first the literature focusing on the differences between novices and experts, followed by literature focusing on the continuity between them at the level of non-propositional knowledge structures and processes, and finally, turn to recent literature beginning to examine the coordination of propositional and non-propositional knowledge.

For a number of decades, there has been an interest in how the problem-solving strategies of experts and novices differ, with many studies focusing on physics problem-solving in particular (e.g. Chi, 2006a; Chi et al., 1981, 1982). While many cognitive scientists have been motivated primarily by interest in foundational understanding of the nature of expertise, studying scientific reasoning by experts has been seen by some as a way to understand how we can help others to become more skilled and knowledgeable, that is, this research has been conducted from an educational perspective (Chi, 2006b; Clement, 2009). Based on Hoffman's (1998) scale of proficiency, Chi (2006b) argues for a continuum of levels from the novice to the master, a person regarded by other experts as being *the* expert. Intermediate levels, paralleling the historical hierarchy of workmanship, include the apprentice, a student engaged in instruction beyond the introductory level, and the journeyman, who has developed a higher level of competence and independence. Early work by Chi et al. (1981, 1982) found that one qualitative difference between novices and experts lies in how they categorise and represent physics problems. Chi et al. (1981) found that novices categorised problems in terms of concrete 'surface similarities', such as the shared physical objects involved in the problem situation (e.g. blocks on an inclined plane), whereas experts categorised problems in terms of more general 'physical principles' (e.g. the law of conservation of energy). Such differences in categorisation may be due to the experts' representation of the problems in terms of idealised objects and the physics concepts involved, such as force, while novices rely on representation of the literal objects and their spatial relationships.

In contrast to Chi et al.'s (1981) focus on differences between novices and experts in the use of propositionally expressed laws and principles, a number of researchers have emphasised the important role that non-propositional knowledge structures and modes of reasoning play in scientific expertise and the continuity that this implies with the cognitive resources available to novices. diSessa (1993) provided an extended account of the role of abstractions from sensorimotor experiences, what he called 'p-prims', in scientific understanding and reasoning. In that account, scientific expertise involved the coordination of multiple p-prims. Given that p-prims are knowledge structures formed early in life through interactions with the physical world, their use by experts reflect a source of continuity between novices and experts. On this view, learning involves a reorganisation of knowledge structures already possessed by the novice. In addition, Clement (2009) has shown that experts make use of non-propositional resources in their approach to solving more challenging physics problems. When he analysed transcripts of scientists' thinking aloud while they solved problems, Clement found that abstract scientific understanding and reasoning are grounded in more concrete, bodily-based knowledge structures and non-propositional modes of reasoning, including analogical reasoning, imagistic simulation and application of physical intuition. These types of knowledge structures and modes of reasoning are assumed to be resources available to the learner and contribute to creative leaps in understanding, and so Clement suggested that these should, therefore, be given more attention in science education. In a similar vein, within cognitive science research, but recognised as applicable to science education (Stolpe & Björklund, 2011), Dreyfus, Dreyfus, and Athanasiou (1986) describe how a person may advance from a novice to expert passing through five levels. They claim that engagement of intuitive reasoning is the crucial feature that distinguishes the two highest levels—'proficiency' and 'expertise'—from the preceding 'novice', 'advanced beginner' and 'competent person' levels, which rely on the adherence to explicit propositional rules.

This interest in the role of non-propositional structures and modes of reasoning in scientific expertise, parallel a more general 'embodiment turn' in cognitive science. Cognitive science has traditionally relied on the analogy of the mind to a computer, and that our cognition can be modelled fruitfully in terms of propositional representations made up of arbitrary symbols and processing that involves the manipulation of those symbols modelled as formal logic (e.g. Larkin, McDermott, Simon, & Simon, 1980). *Embodied cognition* has evolved as a diverse movement, which nonetheless unites around the critique of this traditional approach to cognitive science (e.g. Barsalou, 2008; Clark, 2008; Dreyfus et al., 1986; Lakoff & Johnson, 1980, 1999; Varela, Thompson, & Rosch, 1991). One central idea of embodied cognition is that even the representation of apparently abstract conceptual knowledge and abstract reasoning relies heavily on our concrete experiences of perceptual and motor interaction with our surroundings, as well as our capacity for mental, imagistic simulation.

Contributing to the embodiment turn in cognitive science, Lakoff and Johnson (1980, 1999) have put forward the construct of conceptual metaphor, arguing that we typically come to understand abstract concepts by implicit reference to more concrete, visceral experiences. Such concrete experiences, in turn, are organised by use of *image schemata* (Johnson, 1987), preconceptual abstractions from sensorimotor interaction. An example is the CONTAINER image schema, by which we conceive of objects being located inside or outside of a bounded region in space, or going into or out of it. Furthermore, Lakoff and Johnson (1999) describe two different, but related ways in which we organise our understanding of abstract states and changes of state: the Location Event-Structure metaphor and the Object Event-Structure metaphor. In the Location Event-Structure metaphor, states are construed as if they were locations in space (e.g. 'I'm *in* a depression'), changes are construed as movement between locations (e.g. 'I *got out of* my depression') and caused changes as forced movement (e.g. 'His joy of life *pulled me out of* my depression'). With a subtle figure-ground reversal, in the Object Event-Structure metaphor, states are construed as if they were objects/possessions (e.g. 'I *have* a cold') changes of state are construed as the transfer of possessions (e.g. 'I *got* her cold') and caused changes of state are construed as forced transfer of possessions (e.g. 'She *gave* me her cold'). Lakoff and Núñez (2000) argue that even mathematics—one of the most abstract domains of human thought—is embodied, grounded in image-schematic abstractions from sensorimotor experiences.

An embodied cognition perspective has received increasing attention in science education. Andersson (1986) argued that pupils apply the notion of an *experiential gestalt of causation* (Lakoff & Johnson, 1980), a general structure of causation where an agent causes some kind of change in an object, when thinking about a range of scientific topics including phase changes, electric circuits and vision. Adopting an embodied cognition perspective, Reiner (2000) analysed students' collaborative thought experiments regarding the optimal path in a downhill bike race and found that they used a combination of BALANCE and SYMMETRY image schemata. Consequently, Reiner and Gilbert (2000) suggest that bodily knowledge, expressed as image schemata, help us evoke non-propositional images of forces, through which we can make predictions about future events.

Another line of thought that fits within the embodied cognition tradition, is the idea that *empathising* with, or even experiencing *identification* with aspects of natural phenomena, through imaginative mental simulation of a physical situation, contributes to understanding science (Ochs, Gonzales, & Jacoby, 1996; Root-Bernstein, 2002). Root-Bernstein (2002) points out that many prominent scientists have testified to such immediate rapport with natural phenomena. For instance, virologist Jonas Salk (1983, p. 7) reported that: 'I would picture myself as a virus or a cancer cell, for example, and try to sense what it was like to be either and how the immune system would respond.' Similarly, Einstein (1954, p. 226) claimed: 'There is no logical path to these [universal elementary] laws; only intuition, resting on sympathetic understanding of experience, can reach them.' In their analysis of the discourse in a solid state physics group, Ochs et al. (1996) point out that in addition to detached physics-centred language typical of scientific journals, in oral discourse, the physicists also expressed subjective involvement with the phenomena. They not only conceived themselves as experiencing or manipulating the phenomena, but also *identified with* the phenomena and representations of them in graphs that they drew, as expressed in the title quote: 'When I come down I'm in the domain state.' Consequently, Root-Bernstein (2002, p. 68) promotes the ability to empathise with phenomena as 'an important tool in the mental arsenal' for science students to adopt.

Science educators who have taken an interest in the role of identification with natural phenomena in science learning include Wilensky and Reisman (2006), who encouraged students to model the behaviour and interaction of individual organisms as a way to understand ecosystems. Furthermore, Scherr et al. (2013) have developed the Energy Theater, in which learners take on the role of energy units, which undergo transfer and transformation as they enact physical scenarios. All of these applications of an embodied cognition perspective to scientific expertise and science learning emphasise the continuity between the novice and expert at the level of non-propositional knowledge structures and reasoning processes.

A number of researchers have begun to investigate scientific expertise as the *coordinated use* of propositional and non-propositional knowledge structures and processes (Amin, 2009; Amin, Jeppsson, Haglund, & Strömdahl, 2012; Cheng & Brown, 2010; diSessa & Sherin, 1998; Georgiou, 2014; Jeppsson et al., 2013; Sherin, 2001; Singh, 2002). diSessa and Sherin (1998) extended diSessa's earlier work on p-prims as a constituent of expert scientific understanding and reasoning, by proposing the construct of a *coordination class*. On their account, a coordination class is a knowledge system (seen as an alternative construct to the notion of 'concept') made up of heterogeneous types of knowledge elements including non-propositional p-prims and mental models as well as propositionally formulated beliefs, scientific laws and principles. Developing expert understanding of a scientific concept is to strategically assemble and apply these knowledge elements in such a way as to enable consistent reasoning across a range of physical situations. In addition, in an effort to bridge between accounts of qualitative and quantitative reasoning in scientific expertise, Sherin (2001) has identified what he calls symbolic forms, knowledge structures that enable the meaningful interpretation of elements of physics equations. Symbolic forms are composed of two components: a 'symbol template' which is a generic pattern of symbols in an equation; and an associated intuitive conceptual schema, which is similar to, and in some cases *is* a p-prim. The conceptual schema interprets patterns of symbols in equations. For example, the generic pattern of symbols, or 'symbol template,' consisting of two terms separated by an equal sign ($\square = \square$), can be interpreted in terms of the conceptual schema of two balancing influences (which is the p-prim *balancing*). According to Sherin, an expert problem solver will make use of many symbolic forms which give meaning to aspects of equations and help in connecting between qualitative understanding of a physical situation and its mathematical representation.

Other researchers have also investigated the connection between propositional and non-propositional structures and processes in scientific expertise. Focusing on problem-solving strategies where intuition fails, Singh (2002) investigated 20 physics professors' problem-solving approaches to an introductory numerical physics task, designed to be of a novel structure and, therefore, difficult to target by gut intuition only. Singh found that in many cases, the professors started their problem-solving by searching for conservation principles, but also different kinds of visualisations and analogies to map the unfamiliar problem to a more familiar domain. In other words, they made use of a combination of principle-based reasoning (Chi et al., 1981) and concrete embodied experiences (Clement, 2009) in their problem-solving approach. This coordination has also been studied in the early stages of the acquisition of scientific expertise. For example, in their investigation of third- and sixth-grade students' explanations of magnetic phenomena, Cheng and Brown (2010) found that the most advanced and robust reasoning relied on a combination of intuitive knowledge and more formal (what they called 'verbal-symbolic') knowledge. Similarly, Georgiou (2014) analysed undergraduate students' work on context-rich thermodynamics problems, from the perspective of 'semantic gravity', the degree to which the meaning of language in a particular instance is dependent on its context. She focused on one problem, where the students were asked to explain the mechanism for why frost may form on the outside of a container of compressed gas when its valve is left open for a while. She found that successful, internally coherent responses relied on coordination of abstract physics principles with more context-dependent features of the problems. Poetically, she names failure through over-reliance on abstract principles, and thereby neglect of the particulars of the studied phenomenon, 'the Icarus effect', metaphorical disconnection from earthly matters. From this perspective, developing expertise in physics entails fathoming a broader range of semantic gravity, connecting general laws to particular circumstances.

Some researchers have begun to characterise the pervasive CMs that are implicit in the language of science and have begun to explore how they are implicated in expertise and its acquisition. Brookes and Etkina (2007, 2009) showed how physicists make frequent use of CMs in areas such as quantum physics and Newtonian mechanics. These CMs are encoded in the language of these domains. Brookes and Etkina argue that exposure to this language sometimes causes students to make erroneous (concrete) ontological categorisations of scientific concepts (Slotta, Chi, & Joram, 1995) because of overly literal interpretations of metaphorical expressions (see also Brookes & Etkina, 2015).

In previous work, we have identified extensive use of CMs to construe abstract physical quantities in thermodynamics: energy in *The Feynman Lectures on Physics* (Amin, 2009), and entropy and the second law of thermodynamics in university textbooks (Amin et al., 2012) and in Ph.D. student problem-solving dialogues (Jeppsson et al., 2013). We have suggested that CM may be added to the list of productive intuitive resources that contribute to science learning. More specifically, we found that when solving problems on entropy, Ph.D. students coordinated multiple CMs together, and in conjunction with, the use of symbolic forms, which supported the alignment of qualitative and quantitative reasoning (Jeppsson et al., 2013). Thus, that previous study identified an example of the use of CMs in the complex coordination between non-propositional and propositional knowledge elements in the context of advanced scientific problem-solving. However, the original purpose of that study was broader; it had the more general goal of characterising the roles of CMs in scientific problem-solving. The present study returns to the think aloud protocols of the two Ph.D. students of that previous study and reanalyses that data with a focus on the role of CM specifically in the coordination of propositional and non-propositional knowledge structures. Moreover, it examines undergraduate students' attempts to solve the same problems tackled by the Ph.D. students. It asks whether the acquisition of expertise in scientific problem-solving involves changes in how CMs are used in scientific problem-solving, specifically with regard to how non-propositional and propositional knowledge structures are coordinated. In summary, the research question we attempt to answer is: How does the use of CM differ in scientific problem-solving at different levels of expertise, specifically with regard to how non-propositional and propositional knowledge structures are coordinated? We use the present qualitative study of the reasoning of two Ph.D. students and two undergraduates to generate data-driven hypotheses about differences in the use of CMs at different levels of expertise. Further research is needed to further validate the hypotheses generated and to extend them to a wider range of problem-solving contexts.

## Method

In this study, two pairs of students—a pair of Ph.D. students in physical chemistry and a pair of undergraduate chemistry students—were presented with three problems, each requiring an understanding and application of the concept of entropy. Each pair was asked to work together while thinking aloud to solve the problems. Verbatim transcripts of their problem-solving sessions were prepared and analysed for what they might reveal about the varying use of CM in problem-solving, with specific focus on the coordination of propositional and non-propositional knowledge structures. In this section, we provide information about the participants in the study, the problems used, and the methods of data collection and analysis.

## Participants

Four participants took part in this study: two Swedish Ph.D. students specialising in physical chemistry and two Swedish undergraduate chemistry students. Along Chi's (2006b) novice–master continuum, the Ph.D. students may be characterised as journeymen, while the undergraduates are apprentices. The two Ph.D. students (henceforth, D1 and D2) were well acquainted with thermodynamics at the postgraduate level. At the time of the study, both of them had approached the end of their doctoral studies and had taken graduate courses on statistical thermodynamics and physical chemistry. D1 focused his research on the study of the interaction energies between surfaces in solvents, while D2 studied statistical physics of nanoclusters. The two undergraduate students (S1 and S2) were halfway through the undergraduate chemistry programme and had taken thermodynamics and statistical mechanics within a course on physical chemistry.

## Problems

The participants were presented with three thermodynamics problems, which required an understanding and application of the concept of entropy, two of which—Problems 1 and 3—are in focus in the current study of coordination between formal and non-formal knowledge resources. We summarise these problems here, while more detailed descriptions of the problems and possible solutions can be found in Jeppsson et al. (2013). In Problem 1, the participants were asked to account for the mechanism that drives the process of freezing water in a beaker that is placed in a freezer. A scientifically appropriate response to this question would recognise that as the water freezes, there is heat transfer from the water to the surroundings. During this process, the entropy of the surroundings increases more than the entropy decrease in the water; this net increase in entropy can be seen as what drives the process forward. Problem 3, in turn, involved establishing the entropy change of an ideal gas undergoing reversible, adiabatic expansion. A scientifically appropriate response would recognise that the entropy remains unchanged, since no heat is exchanged with the surroundings and $S = Q/T = 0$ J/K, due to the reversibility assumption.

## Data Collection and Analysis

The student pairs spent approximately one hour discussing solutions to the three problems, in which the Ph.D. students made 523 dialogue turns and the undergraduates made 238, followed by a joint debriefing session of about 20 minutes. These sessions were video recorded. Verbatim transcripts were later prepared for subsequent analysis. The transcripts were entered into the MAXQDA software that supported the qualitative analysis of the problem-solving sessions. The transcripts where analysed in four phases as follows. The first phase of the analysis involved segmenting the problem-solving protocol for each problem into distinct reasoning episodes and interesting episodes were identified in which the pair made some progress towards their preferred solution. The second phase involved the identification of CMs used in the selected episodes. These episodes were read and re-read repeatedly and candidate sentences that seemed to have metaphorical units were intuitively selected and later brought up in a group discussion among the authors on whether they should be regarded as metaphorical phrases or not. As a way to increase the reliability of metaphor identification, a simplified version of the Metaphor Identification Procedure (MIP) developed by the Pragglejaz Group (2007) was applied to the candidate metaphorical sentences. First, the candidate metaphorical lexical unit(s) in the sentence is(are) identified. Next the contextual meaning of the whole sentence is glossed and the meaning of the lexical unit(s) in context is described. The next step is to decide whether it has a more basic, more concrete meaning in another context. If there is, the researcher has to decide whether the basic meaning contrasts with the contextual meaning of the lexical unit. If it does, the usage is considered to be metaphorical. After applying the MIP in each selected episode in this way, instances of use of CMs were then identified by generalising across different metaphorical units and drawing from the literature on CM. Here is an example of our application of the MIP in our analysis.

Example: ' ... delta S as a function of [T] for a certain delta Q, one gets something like this ... '

Candidate metaphorical lexical units: 'one gets'

Contextual meaning:

(1) Of whole sentence: When applying the formula $dS = dQ/T$ and you have a certain amount of heat exchanged at some temperature the resulting change in entropy that can be calculated is ...

(2) Of candidate metaphorical lexical units 'one gets': The result of the calculation is.

In the example above, 'one gets' means something along the lines of 'the result of the calculation is.' However, 'one gets' has a more literal meaning in terms of reception of a tangible object, transfer of possession. By seeing multiple metaphorical expressions dealing with the metaphorical transfer of possession *into* or *out of* a function and based on knowledge of CMs identified in the literature, we interpret, 'one gets' as reflecting the Function Is A Machine CM, where a function is construed as a machine that takes objects as inputs and giving objects as outputs. Therefore, 'one gets' is treated as a metaphorical unit where the abstract output from a function is construed as a concrete object you can physically manipulate in different ways. Note that 'one gets something like this' comes across as perfectly conventional in the context of calculating functions; its metaphorical origin is identified only through careful analysis.

In the third phase of the analysis, the role that these CMs played in the reasoning of the episode was described. Moreover, this role was described in relation to the use of other knowledge elements in the problem solution. For the purposes of this particular study, we were interested in distinguishing the use of propositional and non-propositional knowledge structures in the reasoning. We limited our attempts to identify non-propositional knowledge structures to two types of structures: image-schemas that can be inferred as the source domains of CMs reflected in metaphorical expressions uttered by the participants; and the conceptual schemas associated with symbolic forms inferred in the analysis of the problem-solving protocol. Once this analysis was carried out for the transcript of each pair, a fourth phase of the analysis involved making comparisons across pairs. Of interest was how the use of CMs differed in the problem solutions of the two pairs, with particular focus on the coordination (or absence thereof) of propositional and non-propositional knowledge structures.

We should note that in our previous analysis of the Ph.D. students' problem-solving protocol, we found frequent metaphorical use of pronouns ('I', 'we', 'you', or 'one'). In this study as well, the application of the MIP revealed the metaphorical use of pronouns. Prompted by prior research by others suggesting that expertise in science involves a high degree of conceptual engagement by scientists with their objects of investigation (e.g. Ochs et al. 1996 reviewed above) the level of engagement was compared across levels of expertise. To do this, CMs reflecting three distinct levels of increasing engagement with the phenomenon were identified: A Problem Solver Is An Owner/Observer Of A System, A Problem Solver Is A Manipulator Of A System, and A Problem Solver Is A System. The frequency of use of each of these CMs in each transcript was determined and profiles of use by undergraduates and Ph.D. students were compared.

## Results

In this section, we first present the results of the analyses conducted on the Ph.D. students' and the undergraduate students' problem-solving discussions separately. In presenting the results of our analyses of each pair's problem-solving, we provide an illustrative excerpt from the transcript of the pair's problem-solving discussion; present the CMs used in the problem solution and then explain the extent and nature of the coordination between propositional and non-propositional knowledge structures. At the end of this section, we then turn to a comparison of the roles of CMs in both pairs' problem-solving approaches, with an emphasis on how they differ with respect to the coordination of propositional and non-propositional knowledge structures and the degree of engagement with the physical systems being considered.

*Analysis of the Ph.D. Students' Problem-Solving*

*Problem 1.* In the first problem, the students were asked what drives the freezing of water in a beaker at a temperature of $0\,^{\circ}\mathrm{C}$ when it is moved to a freezer at $-10\,^{\circ}\mathrm{C}$. The following excerpt presents the central episode in their attempt to solve this problem:

D2: Well, in this case ... er, I guess it's simply that ... if **I take** heat **from** this beaker with water ... and **move over to** the room ... in principle, then ... the partition function in ... for the room will increase ... more than what **I lose** in the beaker, then ...

D1: Uhum ...

D2: Now, let's see ... it's ...

D1: ... the entropy, you mean ... ?

D2: Yes, in principle, that is the entropy ... so ... / ... / delta $S$ as a function of [$T$] for a certain delta $Q$, **one gets** something like this ... / ... / so, maybe ... the lower the temperature, the bigger the entropy **gain you get** if **you move some** heat **into** this system ... so, that means for our heat bath system here ... that if **you move** a small amount of heat **from** the water beaker **out to** the room ... then, the entropy in the room will increase more than the entropy has decreased in the water beaker ... due to the reason that it is colder in the room ...

The words marked in bold in the excerpt above indicate lexical items that reflect the use of implicit CMs. In 'If I take heat from ... and move over to ... ' D2 talks about heat as a substance/possession, which he imagines moving from the considered system to its surroundings. This reflects the use of two CMs, Change of State is Transfer of a Possession (applied to heat) and the metaphorical use of the pronoun 'I', which reflects the CM A Problem Solver Is A Manipulator of A System. The same two metaphors are also used (this time using the pronoun 'you') in 'If you move some heat into this system' and 'if you move a small amount of heat from the water beaker out to the room.' Moreover, the CM Change of State is Transfer of a Possession is used again; this time applied to entropy change in 'more than what I lose in the beaker' and 'the bigger the entropy gain you get ... .' In addition, the 'I' in 'more than what I lose' reflects the CM A Problem Solver Is An Owner Of A System. The phrase 'the bigger the entropy gain you get' also reflects two other implicit CMs: The use of 'get' reflects the A Function Is A Machine CM and the use of the pronoun 'you' reflects the CM The Problem Solver Is Manipulator of a Machine.

Our previous analysis of this reasoning episode, reported in Jeppsson et al. (2013), emphasised the coherence of the source domains in the coordination of these CMs. We noted that the source domains used to construe the qualitative and quantitative aspects of the problem, respectively, fit into coherent images, with alignment between the images—that is, the coherent image of a manipulator of the system moving possessions from one location to another construing the changes of state of parts of the system and surroundings, and the manipulator of the machine providing input and receiving output. We suggested that this coordinated use of CMs in order to align qualitative and quantitative reasoning was a central contribution of CMs to problem-solving. We also noted the coordination of these CMs with the use of the 'prop-' symbolic form (Sherin, 2001) to interpret the formula $dS = dQ/T$ with a focus on temperature as the denominator. This allowed the Ph.D. students to compare the entropy change in the beaker and its surroundings, and conclude that the entropy increase in the (lower temperature) surroundings is greater than the entropy decrease in the (higher temperature) beaker.

What we would like to add here is an explicit consideration of the representational modes of the knowledge structures—propositional and non-propositional—being employed in this reasoning. We must note at the outset that this whole excerpt is, of course, realised in language. Careful analysis could be conducted of the gestures the Ph.D. students used (as carried out by Dreyfus, Gupta, & Redish, 2015) and/or the diagrams they drew (Clement, 2009). However, our interest in CM has led us to focus on verbal realisations of the reasoning, that is, propositional reasoning, formulated using the lexical and syntactic resources of natural language. However, our analysis of elements of the language as reflecting CMs allows us to identify underlying image-schematic source domains—namely substances/possessions, transfer of possessions from one location to another, machines, and input to and output from machines.

In our analysis, we also ask: what are the central propositional representations that play a role in the reasoning? We note first D2's remark 'in principle, then ... the partition function in ... for the room will increase ... more than what I lose in the beaker' as central to the reasoning in this excerpt. In fact, we can see the rest as unpacking the meaning underlying this statement. The partition function is introduced in statistical mechanics, representing the way the energy of a system is distributed across discrete energy levels. Here, at a macroscopic level, the partition function may be seen to roughly represent the entropy. Although D2's expression 'in principle' is perhaps a brief discourse marker, it signifies that in his view his solution must adhere to a foundational principle, the second law of thermodynamics, according to which *the total entropy of the system and its surroundings cannot decrease*. This is a strict adherence to a propositionally expressed law and illustrates principle-based reasoning, noted by Chi et al. (1981) as an important characteristic of expertise. Next, the mathematical formula expressing the relationship between change in entropy, exchange of heat and temperature, $dS = dQ/T$, is another propositional representation that plays a central role in the reasoning. In applying this formula to the water in the beaker and the air in the freezer surrounding the beaker and comparing the values, the second law of thermodynamics frames the particular situation in this problem.

Based on this analysis of the resources made use of in the solution to this problem, we hypothesise the following joint contributions of both the propositional and image-schematic (non-propositional) knowledge structures. The propositionally expressed second law and the mathematical formula $dS = dQ/T$ might be seen as *constraints* on the reasoning carried out. The problem solvers know them and know their status as principles that must be obeyed. The image-schematic structures, reflected in metaphorical verbal expressions, allow for the use of more concrete, cognitive resources to interpret these abstract propositions. Moreover, they allow for the meanings of the two propositions to be coordinated easily. Interpreting entropy change as possession transfer in both cases allows for a common interpretation of the concept of entropy in both. Moreover, as we have noted above, interpreting heat as a possession that is transferred at both the qualitative and quantitative levels also allows for coordination between the interpretation of the physical situation being considered, and the mathematical formula being used to draw conclusions about it.

We must acknowledge that this analysis can only be seen as a hypothesised interpretation of descriptive data. For example, the counterargument that the CMs reflected in the transcript are relatively superficial linguistic phenomena not contributing to the reasoning must be entertained. We would argue that the consistent and coherent use of source domains of possession and transfer of possession suggest that what we are revealing is a *conceptual* phenomenon, but other methods (e.g. psycholinguistic techniques and gesture analysis) would be required to adjudicate this. Moreover, the plausibility of treating the propositionally expressed second law of thermodynamics and the mathematical formula $dS = dQ/T$ as 'constraints' on the non-propositional knowledge elements used warrants discussion. Hypothesising this constraint role amounts to making a causal claim—namely that the propositional representation *influences* what non-propositional knowledge elements are triggered. Again, a descriptive study cannot provide unequivocal evidence for this claim. However, suggestive discursive evidence can be provided: the use (twice) of the discourse marker 'in principle;' the appeal to the proposition early in the reasoning episode; and the central role of the CM A Function Is A Machine (with respect to which the other CMs cohere as explained above), which interprets that mathematical formula used to solve the problem.

*Problem 3.* Problem 3 required determining the entropy change of an ideal gas undergoing reversible, adiabatic expansion. As mentioned, the expected answer is that the entropy remains unchanged, since no heat is exchanged with the surroundings and $S = Q/T = 0$ J/K in the reversible process. The following excerpt from the Ph.D. students' problem-solving session is the central episode in their thinking through a solution to Problem 3:

> D2: So, the definition of a reversible process actually is that the entropy does not change ...
> D1: Well, right ... [draws a PV diagram] It is that ... it's a question of that **one walks along the same line** ... if **one** increases the volume ... and then, when **one** decreases the volume, then ...
> D2: ... **you can get back to** the same state ...
> D1: Yes, right.
> D2: Then you can't have had any entropy **losses** ... because **you** can never decrease the entropy in an isolated system ...
> D1: No.
> D2: Because if **you are going to be able to get back to the same point**, then you can't increase it either, right, because then **you won't get back** ...
> D1: It's always strange to think that [the entropy] is the same ... but, well ... I guess that's what it is ... it [the problem-solving approach] goes straight to the entropy ... that it would be presupposed that **one gets more locations to be in** ...

Again, we begin by pointing out the CMs used in this episode. First, there is consistent and joint use of the CMs A Problem Solver Is A System, States (Of A System) Are Locations and Change Of State Is Movement, reflected in the metaphorical use of the pronouns 'one' and 'you' in the phrases 'one walks along the same line' and 'you won't get back', and 'one gets more locations to be in'. As we noted in Jeppsson et al. (2013), this use of pronouns reflects integration of three things: the system, the points on the graph and the problem solver. A different metaphorical use of pronouns can be found in 'you can never decrease the entropy in an isolated system', which reflects the CM A Problem Solver Is A Manipulator of A System. Moreover, we find the CM Change of State is Transfer of a Possession (applied to entropy) in 'entropy losses'.[1]

Next we consider the role of propositional and non-propositional knowledge structures in this reasoning episode. We note first the opening statement made by D2: 'the definition of a reversible process actually is that the entropy does not change.' Again, this reflects the use of a propositional formulation of a foundational characteristic (a 'definition' as D2 puts it) of reversible processes that the total entropy of a system and its surroundings remains unchanged as a constraint on the reasoning that follows (Chi et al., 1981). The use of the CMs and the image-schematic source domain of which they are constituted play a different role in this episode than we saw in the excerpt for Problem 1. In that case, multiple CMs were used to align the construal of aspects of the qualitative and quantitative reasoning, where the latter was central to the solution of the problem. In the case of the excerpt under consideration here, the complex cognitive task of applying the principle of no entropy change in the reversible process to the relationship between the pressure and volume of the gas and the graphical representation of that relationship was supported through the coordination of CMs listed above. Specifically, the constraint implied by the principle seems to be reinterpreted for the purposes of reasoning about this particular problem situation by conceiving of an individual walking along a constrained path with that image interpreting the continuous curve on the pressure–volume graph.

Even though the Ph.D. students had produced an accurate solution to the given problem through a combination of attention to the particulars of the situation and principle-based reasoning, D1 still felt that their answer went against his intuition, as reflected in the last turn of the excerpt: 'It's always strange … ' This discomfort probably stems from the salience of the increase in microstates due to increased volume, in comparison to the negative contribution from the decreased energy to be distributed across the particles. This interpretation is supported by the finding that it is common for students to ascribe an entropy increase to a system undergoing adiabatic, reversible expansion, due to the volume increase, and ignore the reduction of the internal energy of the system (Brosseau & Viard, 1992; Haglund & Jeppsson, 2014). The difficulty of carrying out conventional, scientifically sanctioned metaphorical mappings of intuitive, image-schematic structures to abstract scientific concepts has already been noted in the literature (Brookes & Etkina, 2007). Nevertheless, of particular interest here is that D1 seems to accept D2's principle-based approach that 'goes straight to the entropy' as a constraint, and thus suppresses his intuition (i.e. the more obvious metaphorical mapping) in favour of the propositionally formulated principle.

However, again, it is important to be cautious about such interpretations from descriptive data. Indeed, a potential counterargument to the role of propositional constraints is suggested by an exchange in the debriefing session that followed, where the two students reflected upon the problem-solving process:

> D2: I tested both ... like, microscopic, macroscopic ... do a bit of calculations ... to see what one gets ... so I guess ... one uses the method that seems most straightforward ... / ... / ... to get to an answer.
> D1: One suspects an answer ... and then one works towards it. [laughter]
> D2: Yes, one often has a hunch about ... maybe from a macroscopic perspective, of what the answer should be ... / ... / And then, one can, from a microscopic perspective ... work forward and see if one ... if one gets it [the answer] ... if one doesn't get it, one had probably done something wrong somewhere! [laughter]

Here, the Ph.D. students conceptualise the problem-solving process by recourse to a metaphor of physical travelling, the CM Problem Solving Is Walking Along A Path. This is indicated by expressions, such as 'straightforward', 'get to an answer', 'one works towards it' and 'work forward'. This metaphor referring to travelling is interlaced with another common construal, A Function Is A Machine, as expressed in 'to see what one gets', where the answer is represented as an object obtained from the process rather than a location at which one arrives at the end of the process. The Ph.D. students' expression of suspicion, or having a hunch, suggests that they may begin with the intuitive, qualitative assessment of what might be a reasonable answer followed by deciding on a suitable quantitative approach to confirm it (Dreyfus et al., 1986). However, D2 refers to 'do a bit of calculations' at an early stage of the problem-solving process. So the process of coming up with a hunch may, as we suggested in our interpretation of the Ph.D. students' work with Problem 3, be constrained by identification of relevant physics principles (Chi et al., 1981). What seems quite clear, however, is that successful problem-solving involves a coordination of formal and non-formal knowledge resources.


*Analysis of the Undergraduate Students' Problem-Solving*

We have now seen how the Ph.D. students successfully coordinated different non-propositional, embodied resources with each other and in conjunction with propositionally formulated physical principles. We turn now to investigating how the pair of undergraduate students, apprentices in Chi's (2006b) scheme, approached these problems and what roles propositional and non-propositional knowledge structures play.

*Problem 1.* The undergraduate students started with Problem 1, where they were asked to explain what drives forward the process of freezing water placed in a freezer. The following excerpts reflect the heart of their attempts to reason through this problem:

> S2: Well, what **drives** it all has to be, in some way, that the entropy increases, as always ... / ... /
> S1: The standard solution ... in what way the entropy increases is then ... it's because ... temperature equalisation is always ... there is without exception always an entropy increase [they giggle].

After some time, when the students wrestle with focusing either on the process of freezing of the water or the exchange of heat with the surrounding air, they turn to modelling at the microscopic level:

S1: Well, the energy that gets **released** when … the molecules solidify into crystals … is **used** to heat up a 10 degree colder environment.
S2: Yes. And what is really the thing with freezing, anyway? Or why does it happen at a certain temperature … ? Well, I don't know …

We note at the outset the rather limited use of CMs (with metaphorically used words indicated in bold). The use of 'drives' (which is repeated from the initial formulation of the problem presented to them) reflects the CM Change Of State Is Forced Movement. 'The energy that gets released … ' reflects the CM Change Of State Is Transfer Of Possession, with 'used' elaborating the construal of the possession/energy as a resource. So how are these undergraduate students using propositional and non-propositional knowledge structures in these excerpts? First, they begin, like the Ph.D. students, by stating the propositionally expressed principle that the total entropy of a system and its surroundings increases in processes involving temperature equalisation, that is, the second law of thermodynamics. Their metaphorical use of 'drives' is embedded in this proposition in a conventional, textbook formulation of the second law. Then, the students try to make sense of what is happening at the molecular level in order to provide a mechanistic explanation of the phenomenon (cf. Brown & Clement, 1989). S1 uses the CMs Change of State is Transfer of a Possession (applied to energy) and the metaphorical construal of energy as a resource. Again, these two metaphorical construals of energy are highly conventional metaphorical expressions that the students would have encountered in textbooks. Importantly, their use of metaphors is not productive here (i.e. their use does not lead them to generate novel metaphorical expressions consistent with the underlying mapping as in 'if you take heat from the beaker and move to … ' from the Ph.D. students' problem-solving described above). In addition, their use does not help them interpret the law that they state at the beginning or link it to the specific physical situation they are considering in the problem; yet another case of 'the Icarus effect' (Georgiou, 2014). Moreover, they are not able to retrieve the appropriate mathematical representation ($dS = dQ/T$) that could help them develop their answer. In the absence of these, their problem-solving gets stuck and they do not manage to connect the stated principle of entropy increase with the mechanism of heat transfer from the system to its surroundings. Instead, S2 ends up puzzled with the role of water freezing at a certain temperature.

*Problem 3.* We turn next to the undergraduates' dialogue regarding Problem 3. The central reasoning episode for this problem is presented in the following excerpt:

S1: But this is … here, **you** do not have any heat **exchange** with the surroundings … and since I did this quite recently, I remember that the entropy … I mean, the entropy change of an adiabatic process is zero … and here it is, because Q [the exchanged heat] is equal to zero … / … /
S2: … or the entropy increase when the gas expands is **counteracted** by there being a temperature decrease … / … / … well, temperature decrease is that every molecule **gets** less kinetic energy. / … /
S1: When **one** pulls out … or it [the piston] is pushed out like this … and the molecules collide with the wall and exert pressure … they **lose** energy, because it is on its way outwards. / … / so then, they **have** lower velocity when they fly back from the collisions. It may seem very strange! The volume increases enormously, but the entropy does not increase and it gets a lot colder.

Marked in bold are words used metaphorically reflecting the use of a number of CMs. The CM Change of State is Transfer of Possession is applied to heat (reflected in 'heat exchange') and energy ('every molecule gets less kinetic energy' and 'they lose energy'). In addition, the metaphorical use of pronouns indicates two underlying CMs: A Problem Solver Is An Owner of A System (reflected in 'you do not have any heat exchange') and A Problem Solver Is A Manipulator of A System (reflected in 'when one pulls out [the piston]').

We now focus on, as with the previous episodes analysed, the roles of propositional and non-propositional knowledge structures. S1 recalls having encountered a similar situation/problem recently, and in line with the Ph.D. students, she begins by stating the propositionally formulated principle that adiabatic processes do not involve heat exchange, which—in combination with the stated reversibility of the process—leads to a zero entropy change. It is reasonable to infer that this conclusion relies on the implicit assumption of the (propositional) mathematical formula $dS = dQ/T = 0$ J/K. These two propositions help the pair get straight to an answer of the entropy change, and this seems to constrain the reasoning that follows as they explain the underlying physical process. (Again, the hypothesis that these propositions play the role of constraint on reasoning comes from the discourse, for example, an authoritative statement with no hedging—'the entropy change of an adiabatic process *is* zero'— and the subsequent reasoning assumes total entropy change is zero.)

The relatively few CMs are used in the following way. The CM Change Of State Is Transfer Of Possession (applied to heat) is a direct (conventional) metaphorical use based on knowing what adiabatic means, expressed in 'you do not have any heat exchange with the surroundings'. This supports the interpretation of the implicitly invoked formula $dS = dQ/T$. When providing their explanation of the process, the entropy change is construed as consisting of two components—associated with the increase in volume (which contributes positively to the entropy change) and the reduction in temperature (which contributes negatively to the entropy change). The use of 'counteracts' is a metaphorical construal of these two contributions as two opposing tendencies, interpreted in terms of force dynamics (Talmy, 1988). Similar to the Ph.D. students, S1 finds the unchanged entropy 'strange', that is, not in line with her immediate intuitions, and this tension between the result of their principle-based reasoning and intuition prompts them to search for a way to reconcile the ideas.

S2 introduces a microscopic perspective in connecting the temperature decrease to the average kinetic energy of the molecules of the gas. S1 adopts this molecular perspective in providing an explanatory model (Brown & Clement, 1989), where she imagines what would happen 'when one pulls out' the piston, by using the CM A Problem Solver Is A Manipulator Of A System. The CMs Change Of State As Transfer Of Possession (applied to energy) and the States Are Possessions (applied to velocity) are also used in this explanation, which involves the imagistic simulation of the situation with conventional metaphorical interpretations of energy and velocity. However, as opposed to the Ph.D. students' expression 'one walks along the same line' in relation to the same problem, in this imagistic reasoning, there is a clear distinction between the undergraduate students themselves as manipulators and the molecules as components of the system, referred to as 'they' in a detached way. Nonetheless, these undergraduates, with one of them having just encountered the problem before, manage to coordinate propositional principles at the macroscopic level with imagistic reasoning, and are able to make sense of the propositionally expressed principle at the microscopic level.

### Contrasting Ph.D. and Undergraduate Student Problem-Solving

We have now seen examples of how the Ph.D. and undergraduate students approached the exercises, and how successful problem-solving largely depended on a productive coordination of appropriate physics principles and non-propositional imagistic resources. The analysis shows that the Ph.D. students are more likely to use CMs to interpret aspects of propositional representations and to coordinate joint use of multiple propositional representations. Moreover, they are more likely to use CMs to help apply propositional representations to specific problem situations. Partial success at problem-solving on the part of the undergraduate students indicates some joint use of propositional and non-propositional representation. However, CMs were minimally used to interpret aspects of propositional representations and relate propositions to the physical situation. Instead, they were used to develop an explanatory model, that might help make sense of the correctness of the proposition.

An additional dimension along which the Ph.D. students differ from the undergraduates is the degree to which they engage or empathise with the studied phenomena (Ochs et al., 1996; Root-Bernstein, 2002). Even though both pairs engage with the modelled systems and manage to coordinate principles and intuitive resources in Problem 3, where the Ph.D. students imagine how 'one walks along the same line' in a very visceral way, the undergraduate students are still rather confined to formulaic physics language, as in 'you do not have any heat exchange with the surroundings'. In fact, the detached language of the undergraduate students often comes across as similar to formal written science texts.

This difference in the reasoning and dialogue between the pairs is more a matter of degree than a clear dichotomy (which might reflect the closeness of the apprentice and journeyman categories in Hoffman's (1998) hierarchy of levels of expertise). Nevertheless, a quantitative pattern emerges from the way the two pairs used CMs, reflecting their own role as problem solvers, in combination with personal pronouns. This is concluded from an analysis of the metaphorical use of pronouns in all dialogue turns of the problem-solving exercises of the two pairs (see Table 1), using a categorisation scheme of three levels of engagement, inspired by Ochs et al. (1996). At the lowest level of engagement, the problem solver conceives of observing or owning the considered system, but does not imagine interacting with it. At the next level, the problem solver envisions manipulating the system, and thinks through the consequences. Finally, at the highest level of engagement, the problem solver actually identifies with the system.

Overall, the Ph.D.s provided considerably more metaphorical construals of this type than the undergraduates (51 compared to 10, respectively), which cannot be explained

Table 1. Frequencies of pronouns ('I', 'you', 'we', or 'one') used as constituents in CMs, which reflect different levels of engagement with the phenomena considered in the problem-solving dialogues

| CMs | Undergraduates | Ph.D.s | Example utterance |
|---|---|---|---|
| A Problem Solver Is An Owner/ Observer Of A System | 1 | 30 | 'more than what I lose in the beaker' (Ph.D.s) |
| A Problem Solver Is A Manipulator Of A System | 9 | 17 | 'when one pulls out [the piston]' (undergraduates) |
| A Problem Solver Is A System | 0 | 4 | 'one walks along the same line' (Ph.D.s) |

solely by their larger number of turns overall (523 vs. 238). It may be noted that the most frequent category for the Ph.D. students, as opposed to the undergraduates, was the lowest level of engagement, that of an observer or owner of the system. Nevertheless, some of these utterances, such as 'more than what I lose in the beaker' are far from typical formal science text, and could plausibly have been coded as cases of identification with the system, rather than that of an owner of the system (but we conservatively decided on interpretations of less engagement when in doubt). In addition, only the Ph.D. students arrived at the highest level of engagement, identification, as in: 'it's a question of that one walks along the same line'.

## Discussion

*CMs and the Coordination of Propositional and Non-propositional Knowledge Structures in Problem-Solving*

The present study sought to contribute to the existing literature on development of expertise in scientific problem-solving. Specifically, it intended to add to that body of literature that has identified the coordination of propositional and non-propositional knowledge structures by focusing on the role of CMs in this coordination at two different levels of expertise: apprentices (undergraduate students) and journeymen (Ph.D. students). As summarised at the end of the last section, we found that journeymen invoked propositionally expressed principles and laws that were relevant for any given problem, which then served as constraints on their formulation and elaboration of a problem solution. Non-propositional knowledge structures in the form of the image-schematic source domains of CMs and the conceptual schemas of symbolic forms were then invoked and coordinated to interpret these propositional structures. A particularly prominent aspect of the use of CMs by the journeymen was the degree of engagement between the problem solvers, on the one hand, and the physical situation and quantitative reasoning, on the other. The use of CMs transformed what might have been expected to be highly formal reasoning to a process of reasoning that contained many elaborate concrete, imagistic scenarios in which the problem solver himself/herself is construed as a component. In contrast, while often able to invoke the appropriate law or principle, the apprentices were either unable to coordinate, or were limited in coordinating, the needed imagistic elements that would support adequate interpretation of the propositions and laws, apply them to the physical situations or align qualitative and quantitative reasoning.

In light of the analyses presented here, we hypothesise that CMs play a subtle role in the development of expertise in scientific problem-solving. We suggest that they serve as flexible resources that allow the problem solver to construe abstract concepts in terms of a variety of more concrete schemata and serve to coordinate understanding of verbally formulated scientific principles, mathematically expressed laws and concrete images of the physical situations being reasoned about. As we have seen, diSessa and Sherin (1998) have described this kind of dynamic knowledge system in terms of a coordination class composed of multiple elements of different kinds: propositionally expressed beliefs, image schemas (e.g. p-prims) and mental models. The role of such a system is to support the ability to first 'read out' (or 'see') certain physical quantities in particular situations, and then to infer various other quantities and their magnitudes. Image schematic structures are seen as playing an important role for both reading out and inferring, with propositional structures playing a particularly important role in the latter. Thus, we suggest that CMs are potentially important components of coordination classes. Future research is still needed to support this claim further and to clarify the nature of the role of CMs as a component of coordination classes alongside other knowledge elements.

The present study also adds to previous work on the coordination of propositional and non-propositional knowledge structures (Cheng & Brown, 2010; Georgiou, 2014; Singh, 2002) by pointing out the role of a knowledge type that has not been studied much to date—that is, CMs—and by being more explicit about the nature and role of a variety of knowledge elements involved in expert scientific problem-solving. Moreover, as we have pointed out and discussed in previous work (Amin, 2009; Amin et al., 2012; Jeppsson et al., 2013), the use of concrete image-schematic source domains of CMs in abstract, scientific problem-solving shows that the role of ontological classification in the development of scientific expertise is not at all simple. In light of the present study, we can add that implicit concrete construals of abstract concepts such as entropy, energy and heat in terms of possessions, movement of possession and containment in the context of advanced scientific problem-solving can be more prevalent at higher levels of expertise, and not necessarily a sign of naïve reasoning (Chi, Slotta, & De Leeuw, 1994). This makes more complex the question of what ontological shifts occur with the development of expertise.

*Indication of Coordination of Principles and Non-formal Resources in a Master's Formal Science Writing*

In this study, we have focused our analysis on problem-solving dialogue, performed by apprentices and journeymen, at various points along their journey from novice to master. It may be worth investigating whether the identified patterns extend across other genres of scientific communication or along the novice–expert continuum.

For this purpose, we have also looked at Einstein's (1905/1998) seminal paper where he introduced a quantum interpretation of the photoelectric effect, for which he was awarded the 1921 Nobel Prize in Physics. In Hoffman's (1998) classification along the novice–master continuum, Einstein is undoubtedly a master.

First, it should be noted that the entire paper is framed as an ontological discussion, in terms of what is the fundamental character of energy:

> According to Maxwell's theory, energy is considered to be a continuous spatial function for all purely electromagnetic phenomena, hence also for light, whereas according to the present view of physicists, the energy of a ponderable body should be represented as a sum over the atoms and electrons. (Einstein, 1905/1998, p. 177)

In contrast to Maxwell's view, in this paper, Einstein suggests that energy is quantised, that is, distributed discontinuously, also in relation to electromagnetic phenomena. In other words, from the perspective of Chi's line of work (e.g. Chi et al., 1981, 1994), this is a case where the master of masters is problematising the ontological categorisation of one of the central concepts in science. He challenges the physical principles that had been adopted up until that point—the fundamental premises of classical physics, and thereby opens up for the modern physics of the 1900s. In his account of the photoelectric effect, he says:

> According to the view that the incident light consists of energy quanta of energy $(R/N)\beta v$, the production of cathode rays by light can be conceived in the following way. The body's surface layer is penetrated by energy quanta whose energy is converted at least partly into kinetic energy of the electrons. The simplest conception is that a light quantum transfers its entire energy to a single electron; we will assume that this can occur. However, we will not exclude the possibility that electrons absorb only a part of the energy of the light quanta. (Einstein, 1905/1998, p. 194)

In the first sentence, Einstein restates the main claim of the paper: the physical principle that energy is quantised. Next, against this background, he invites us to imagine the process of production of cathode rays, a concrete physical phenomenon. In this way, he frames the ontological discussion within the particularities of an imagined situation, a case of coordination of formal and non-formal knowledge resources. In spite of this invitation to imaginative thought, in this excerpt, Einstein does not express personal engagement with the phenomenon, of the kind presented by Ochs et al. (1996) and adopted in the problem-solving dialogues of the current study. He does use non-propositional resources in the form of metaphor, however. Einstein describes how energy quanta 'penetrate' a body's surface layer, a process in which some or all of its energy is 'transferred to' and 'converted into' kinetic energy of the electrons. In our interpretation, the energy quanta or light quanta are construed as a kind of 'energy carrier' (Falk, Herrmann, & Schmid, 1983), that collides with and delivers a certain amount of energy to the surface, while energy is treated as an object that moves from one location to another; a case of the use of an Object Event-Structure metaphor in relation to energy. In addition, in 'energy is converted / … / into kinetic energy', different energy forms, such as kinetic energy are construed as locations, into which the energy transforms, a use of the CM Forms Of Energy Are

Locations/Containers (Amin, 2009). Finally, when discussing whether all or only some of the energy of a light quantum is transferred to an individual electron, Einstein writes that the electrons 'absorb' energy, a liquid-like version of the Object Event-Structure metaphor.

It may be worth noting, however, that whereas Einstein coordinates imagistic thought with the postulated physics principle of quantised energy, he uses considerably more detached language in relation to the concept of entropy throughout the paper. For instance, in relating Planck's quantum approach to come to terms with the problem of infinite energy density of black-body radiation, the so-called ultraviolet catastrophe, Einstein (1905/1998, p. 186) says:

> If we restrict ourselves to investigating the dependence of the entropy on the volume occupied by the radiation, and denote the entropy of the radiation by $S_0$ at volume $v_0$, we obtain $S - S_0 = \dfrac{E}{\beta v} \ln \left[ \dfrac{v}{v_0} \right].$

Here, in contrast to the visceral nature of energy as an embodied entity in the excerpts above, entropy is construed as a purely mathematical quantity, the value of which is 'obtained' from algebraic calculations.

In conclusion, even in this formal genre, the scientific paper, a master employs imagistic thought when querying into the fundamental nature of energy. However, the contrasting case of entropy shows that such coordination with imagistic thought is used selectively in relation to different concepts and contexts of explanation.

### Implications for Science Education Research and Practice

The view of developing expertise in scientific problem-solving that emerges from this study has a number of implications for science education. We focus our discussion here on the implications of the complexity of the coordination of knowledge elements required, the role of propositional knowledge structures as constraints on that coordination, and the implicit and subtle nature of the metaphorical construals involved.

The development of expertise in scientific problem-solving seems to involve the complex coordination of various kinds of propositional and non-propositional knowledge structures. Moreover, the specifics of these coordinations are different in the context of different problem situations. An implication of these findings is in line with the general educational implications of a situated view of cognition (Greeno, 1989; Lave, 1988). The development of competence in an academic discipline relies on gaining experiences with authentic *practices*, and not merely a matter of exposure to abstract principles. As acknowledged by Hake (1998, p. 65) in relation to undergraduate mechanics courses, traditional approaches 'relying primarily on passive-student lectures, recipe labs, and algorithmic-problem exams' have not been found to support conceptual understanding among the participants in a sufficient way, but should be replaced by approaches characterised by 'interactive engagement'.

One possible approach for students to get exposed to authentic science practices is different kinds of internship or apprenticeship—in a more literal sense than in the scheme adopted by Chi (2006b) and as used up until now in this study. If students are invited to science laboratories, adopting the role of 'peripheral participants' (Lave, 1988), they get to experience how scientists apply abstract principles to concrete phenomena and problems in authentic discourse. Such direct involvement in science-laboratory activities typically begins towards the end of undergraduate programmes, but might be considered much earlier.

Another approach to encourage interactive engagement and coordination of propositional and non-propositional resources is one that has been developed by Scherr et al. (2013) within their Energy Courses. Here participants are introduced to physical phenomena of increasing complexity, and invited to make sense of the involved concepts and represent the phenomena in different ways in peer-group collaboration.

While much of the skill of scientific problem-solving that we have described in this study is of an implicit, subtle nature, we have shown that invoking appropriate principles and laws and explicitly treating them as constraints on the process of problem-solving was an important contributor to successful problem-solving. This implies an important role for explicit instruction into scientific problem-solving. Instructing learners into identifying suitable principles and laws and treating them as constraints, even when intuitions can suggest otherwise, is an important metacognitive stance to encourage. This is consistent with traditional instruction, but learners must also be given the opportunity to engage in a form of sense making that draws on non-propositional knowledge structures as well. This paper has suggested that we consider CMs among those structures that might be relevant to expertise acquisition.

## Disclosure statement

No potential conflict of interest was reported by the authors.

## Notes

1. The interaction between these three conceptual domains could be analysed from the perspective of 'conceptual blending' (Fauconnier & Turner, 1998). We have not adopted and applied this framework formally here, but see, for example, Close and Scherr, 2015, and Dreyfus et al. (2015), for examples of such analyses.

## References

Amin, T. G. (2009). Conceptual metaphor meets conceptual change. *Human Development*, 52(3), 165–197.
Amin, T. G., Jeppsson, F., Haglund, J., & Strömdahl, H. (2012). The arrow of time: Metaphorical construals of entropy and the second law of thermodynamics. *Science Education*, 96(5), 818–848.

Amin, T. G., Smith, C. L., & Wiser, M. (2014). Student conceptions and conceptual change: Three overlapping phases of research. In N. G. Lederman & S. K. Abell (Eds.), *Handbook of research in science education*, Volume 2 (pp. 57–81). New York, NY: Routledge.

Andersson, B. (1986). The experiential gestalt of causation: A common core to pupils' preconceptions in science. *European Journal of Science Education*, 8(2), 155–171.

Barsalou, L. (2008). Grounded cognition. *Annual Review of Psychology*, 59(1), 617–645.

Brookes, D. T., & Etkina, E. (2007). Using conceptual metaphor and functional grammar to explore how language used in physics affects student learning. *Physical Review Special Topics—Physics Education Research*, 3(1), 010105.

Brookes, D. T., & Etkina, E. (2009). "Force," ontology, and language. *Physical Review Special Topics—Physics Education Research*, 5(1), 010110.

Brookes, D. T., & Etkina, E. (2015). The importance of language in students' reasoning about heat in thermodynamic processes. *International Journal of Science Education*. doi:10.1080/09500693. 2015.1025246

Brosseau, C., & Viard, J. (1992). Quelques réflexions sur le concept d'entropie issues d'un enseignement de thermodynamique [Some reflections on the entropy concept from thermodynamics teaching]. *Enseñanza de las ciencias*, 10(1), 13–16.

Brown, D. E., & Clement, J. (1989). Overcoming misconceptions via analogical reasoning: Abstract transfer versus explanatory model construction. *Instructional Science*, 18(4), 237–261.

Cheng, M. F., & Brown, D. E. (2010). Conceptual resources in self-developed explanatory models: The importance of integrating conscious and intuitive knowledge. *International Journal of Science Education*, 32(17), 2367–2392.

Chi, M. T. H. (2006a). Laboratory methods for assessing experts' and novices' knowledge. In A. Ericsson, N. Charness, P. J. Feltovich, & R. Hoffman (Eds.), *The Cambridge handbook of expertise and expert performance* (pp. 167–184). New York, NY: Cambridge University Press.

Chi, M. T. H. (2006b). Two approaches to the study of experts' characteristics. In A. Ericsson, N. Charness, P. J. Feltovich, & R. Hoffman (Eds.), *The Cambridge handbook of expertise and expert performance* (pp. 21–30). New York, NY: Cambridge University Press.

Chi, M. T. H., Feltovich, P. J., & Glaser, R. (1981). Categorization and representation of physics problems by experts and novices. *Cognitive Science*, 5(2), 121–152.

Chi, M. T. H., Glaser, R., & Rees, E. (1982). Expertise in problem solving. In R. J. Sternberg (Ed.), *Advances in the psychology of human intelligence* (Vol. 1, pp. 7–75). Hillsdale, NJ: Erlbaum.

Chi, M. T. H., Slotta, J. D., & De Leeuw, N. (1994). From things to processes: A theory of conceptual change for learning science concepts. *Learning and Instruction*, 4(1), 27–43.

Clark, A. (2008). *Supersizing the mind: Embodiment, action, and cognitive extension*. Oxford: Oxford University Press.

Clement, J. (2009). *Creative model construction in scientists and students: The role of imagery, analogy, and mental simulation*. Dordrecht, NL: Springer.

Close, H., & Scherr, R. (2015). Enacting conceptual metaphor through blending: Learning activities embodying the substance metaphor for energy. *International Journal of Science Education*. doi:10.1080/09500693.2015.1025307

diSessa, A. A. (1993). Toward an epistemology of physics. *Cognition and Instruction*, 10(2–3), 105–225.

diSessa, A. A., & Sherin, B. L. (1998). What changes in conceptual change? *International Journal of Science Education*, 20(10), 1155–1191.

Dreyfus, H. L., Dreyfus, S. E., & Athanasiou, T. (1986). *Mind over machine: The power of human intuition and expertise in the era of the computer*. New York, NY: Free Press.

Dreyfus, B., Gupta, A., & Redish, E. (2015). Applying conceptual blending to model coordinated use of multiple ontological metaphors. *International Journal of Science Education*. doi:10.1080/09500693.2015.1025306

Einstein, A. (1905/1998). On a heuristic point of view concerning the production and transformation of light. In J. Stachel (Ed.), *Einstein's miraculous year: Five papers that changed the world of physics* (pp. 177–198). Princeton, NJ: Princeton University Press.

Einstein, A. (1954). *Ideas and opinions*. New York, NY: Crown.

Falk, G., Herrmann, F., & Schmid, G. B. (1983). Energy forms or energy carriers? *American Journal of Physics*, *51*(12), 1074–1077.

Fauconnier, G., & Turner, M. (1998). Conceptual integration networks. *Cognitive Science*, *22*(2), 133–187.

Georgiou, H. (2014). *Doing positive work: On student understanding of thermodynamics* (PhD dissertation), The University of Sydney, Sydney, Australia.

Greeno, J. G. (1989). A perspective on thinking. *American Psychologist*, *44*(2), 134–141.

Haglund, J., & Jeppsson, F. (2014). Confronting conceptual challenges in thermodynamics by use of self-generated analogies. *Science & Education*, *23*(7), 1505–1529.

Hake, R. R. (1998). Interactive-engagement versus traditional methods: A six-thousand-student survey of mechanics test data for introductory physics courses. *American Journal of Physics*, *66*(1), 64–74.

Hammer, D. (2000). Student resources for learning introductory physics. *American Journal of Physics*, *68*(S1), S52–S59.

Hoffman, R. R. (1998). How can expertise be defined? Implications of research from cognitive psychology. In R. Williams, W. Faulkner, & J. Fleck (Eds.), *Exploring expertise* (pp. 81–100). Mahwah, NJ: Erlbaum.

Jeppsson, F., Haglund, J., Amin, T. G., & Strömdahl, H. (2013). Exploring the use of conceptual metaphors in solving problems on entropy. *Journal of the Learning Sciences*, *22*(1), 70–120.

Johnson, M. (1987). *The body in the mind: The bodily basis of meaning, imagination, and reason*. Chicago, IL: University of Chicago Press.

Lakoff, G., & Johnson, M. (1980). *Metaphors we live by*. Chicago, IL: The University of Chicago Press.

Lakoff, G., & Johnson, M. (1999). *Philosophy in the flesh*. New York, NY: Basic Books.

Lakoff, G., & Núñez, R. E. (2000). *Where mathematics comes from: How the embodied mind brings mathematics into being*. New York, NY: Basic Books.

Larkin, J., McDermott, J., Simon, D. P., & Simon, H. A. (1980). Expert and novice performance in solving physics problems. *Science*, *208*(4450), 1335–1342.

Lave, J. (1988). *Cognition in practice: Mind, mathematics and culture in everyday life*. Cambridge: Cambridge University Press.

Ochs, E., Gonzales, P., & Jacoby, S. (1996). "When I come down I'm in the domain state": Grammar and graphic representation in the interpretive activity of physicists. In E. Ochs, E. A. Schegloff, & S. A. Thompson (Eds.), *Interaction and grammar* (pp. 328–369). Cambridge: Cambridge University Press.

Pragglejaz Group. (2007). MIP: A method for identifying metaphorically used words in discourse. *Metaphor and Symbol*, *22*(1), 1–39.

Reiner, M. (2000). Thought experiments and embodied cognition. In J. K. Gilbert & C. J. Boulter (Eds.), *Developing models in science education* (pp. 157–176). Dordrecht, the Netherlands: Kluwer.

Reiner, M., & Gilbert, J. (2000). Epistemological resources for thought experimentation in science learning. *International Journal of Science Education*, *22*(5), 489–506.

Root-Bernstein, R. S. (2002). Aesthetic cognition. *International Studies in the Philosophy of Science*, *16*(1), 61–77.

Salk, J. (1983). *The anatomy of reality*. New York, NY: Columbia University Press.

Scherr, R. E., Close, H. G., Close, E. W., Flood, V. J., McKagan, S. B., Robertson, A. D., Wittman, M. C., & Vokos, S. (2013). Negotiating energy dynamics through embodied action in a materially structured environment. *Physical Review Special Topics—Physics Education Research, 9*(2), 020105.

Sherin, B. L. (2001). How students understand physics equations. *Cognition and Instruction, 19*(4), 479–541.

Sherin, B. L. (2006). Common sense clarified: The role of intuitive knowledge in physics problem solving. *Journal of Research in Science Teaching, 43*(6), 535–555.

Singh, C. (2002). When physical intuition fails. *American Journal of Physics, 70*(11), 1103–1109.

Slotta, J. D., Chi, M. T. H., & Joram, E. (1995). Assessing students' misclassifications of physics concepts: An ontological basis for conceptual change. *Cognition and Instruction, 13*(3), 373–400.

Smith, J. P., diSessa, A. A., & Roschelle, J. (1993). Misconceptions reconceived: A constructivist analysis of knowledge in transition. *Journal of the Learning Sciences, 3*(2), 115–163.

Stolpe, K., & Björklund, L. (2011). Seeing the wood for the trees: Applying the dual-memory system model to investigate expert teachers' observational skills in natural ecological learning environments. *International Journal of Science Education, 34*(1), 101–125.

Talmy, L. (1988). Force dynamics in language and cognition. *Cognitive Science, 12*(1), 49–100.

Varela, F. J., Thompson, E., & Rosch, E. (1991). *The embodied mind: Cognitive science and human experience.* Cambridge: MIT Press.

Wilensky, U., & Reisman, K. (2006). Thinking like a wolf, a sheep, or a firefly: Learning biology through constructing and testing computational theories—An embodied modeling approach. *Cognition and Instruction, 24*(2), 171–209.

# On Conceptual Metaphor and the Flora and Fauna of Mind: Commentary on Brookes and Etkina; and Jeppsson, Haglund, and Amin

Bruce Sherin

*School of Education and Social Policy, Northwestern University, Evanston, IL, USA*

I have been asked to comment on two papers appearing in this special issue. The first paper is *The Importance of Language in Students' Reasoning About Heat in Thermodynamic Processes*, by David T. Brookes and Eugenia Etkina. The second is *Varying Use of Conceptual Metaphors Across Levels of Expertise in Thermodynamics*, by Fredrik Jeppsson, Jesper Haglund, and Tamer G. Amin.

I will begin with my top-level comment: I found both of these papers to be very congenial, in the sense that their overall thrust is in line with my own core beliefs about how science is learned, and how it should be taught. These core beliefs that I share with the authors are—to use my own vocabulary—that (1) science learning must build on everyday resources that students possess prior to formal science instruction and that (2) successful science learning requires the coordination of these everyday resources and new resources acquired during formal instruction. The authors of both papers are very clear that they share these core beliefs with a range of prior work that includes, in particular, the work of diSessa (1993), as well as my own work on symbolic forms, which they graciously cite (Sherin, 2001).

If these papers are in such good alignment with work that is over 20 years old, then that raises the obvious question: What's new here? Are there ways in which this newer work adds to, or even departs from, for example, the program laid out in diSessa (1993). The short answer is that there are certainly substantial novel contributions in these papers. I cannot fully summarize these contributions, given the small space allotted to me, so I will simply highlight what is most exciting to me.

Over the past 20–30 years, our field has seen substantial research that documents how less expert students understand (and misunderstand) science. However, there has been much less work that examines advanced populations, particularly as they wrestle with truly difficult subject matter. The work described in diSessa (1993), for example, was based on interviews with novice physics students, working only on relatively elementary subject matter, though diSessa speculates about the trajectories from novice to expertise.

In contrast, the work in these papers begins to flesh out the longer term trajectory of science learning. They look at more expert participants, working on notoriously difficult subject matter—the thermodynamic notions of heat and entropy. The authors of these papers make a compelling case that the development of expertise in physics subject matter requires the weaving together—coordination—of old and new knowledge resources. With relatively few exceptions, there has been little research that makes this case in such a compelling way (for one of these exceptions, see Clement, 1994). It is worth pointing out that there are still scholars in our field who argue that there are scientific conceptions, such as heat and entropy, that depart so much from our everyday conceptual apparatus, that their understanding cannot be built on existing conceptual resources (Chi, 2005).

To this point, I have been painting a picture in which the work described in this program can be seen as a logical continuation of work begun decades ago. However, the authors of these papers do not see themselves as merely fleshing out a research program that was begun 20 years ago by diSessa and others. Instead, they see themselves as advancing a new perspective, one that aligns with the themes of this special issue, conceptual metaphor and embodied cognition. Furthermore, I believe that there are some respects in which this new perspective is not compatible with the older program—or, at least with my own take on that older program. In the rest of this paper, I will lay out these incompatibilities, and let readers make of them what they will. To lay my cards on the table, the reader will see that I harbor some skepticism regarding the newer perspective of which embodiment and conceptual metaphor are a part.

## Locating These Papers

I want to start with how the authors position their work in relation to longer term trends in the field. Brookes and Etkina, I believe, see their particular focus on language as something that sets them apart from older work. I agree. It is not that researchers such as myself ignored language; we certainly listened to what participants said. But to a great extent we looked *through* language rather than *at* language. We used the speech of our participants as a window into their thinking, but we did not focus as closely on the window itself.

I take more issue with how Jeppsson et al. position their work with respect to longer term trends in the field. I will start by quoting some text from their introduction:

> Early work on expertise focused on the differences between novices and experts, identifying the important role of abstract principles in the latter, absent in the former (Chi,

2006a, 2006b; Chi, Feltovich, & Glaser, 1981; Chi, Glaser, & Rees, 1982). Another line of work has focused on continuity between novices and experts by focusing on non-propositional knowledge structures such as imagery, image-schemas (abstractions from sensorimotor schemas) and mental models (Clement, 2009; diSessa, 1993; Hammer, 2000; Smith, diSessa, & Roschelle, 1993). This latter approach can be seen as reflecting a kind of "embodiment turn" in research on scientific expertise and science learning. (Jeppsson, Haglund, & Amin, 2015)

Throughout their paper, Jeppsson et al. describe researchers such as Chi as identifying 'propositional' knowledge, and they see this propositional knowledge as aligning with the formal knowledge taught in textbooks. In contrast, they describe work in the diSessa line as non-propositional and embodied.

I want to take issue with some core aspects of these characterizations. First, I do not believe that that we can so straightforwardly say that researchers such as Chi and diSessa are identifying knowledge that differs in *kind*. For example, Chi's problem categories can, I believe, be rightly thought of as problem *schemas*, with all of the potential fuzziness of schemas, such as those associated with p-prims. Second, I do not believe that diSessa and his colleagues (among whom I include myself) would be entirely comfortable with being part of the 'embodiment turn'. (I will say more about this below.) Thus, in sum, aligning non-propositional knowledge with the embodied perspective, and propositional with formal physics knowledge, is just too crude a gloss.

There is a larger issue here, and it is a problem that I think characterizes much of the literature on embodied cognition. Namely, there is a tendency to want to subsume too much into the embodied cognition perspective, and to grant too little to the alternatives. I will quote one other passage from Jeppsson et al. that allows me to make this point dramatically:

> Cognitive science has traditionally relied on the analogy of the mind to a computer, and that our cognition can be modelled fruitfully in terms of propositional representations made up of arbitrary symbols and processing that involves the manipulation of those symbols modelled as formal logic (e.g. Larkin, McDermott, Simon, & Simon, 1980). Embodied cognition has evolved as a diverse movement, which nonetheless unites around the critique of this traditional approach to cognitive science. (Jeppsson et al., 2015)

Even where it relies very heavily on the mind–computer analogy, cognitive science has never been only about propositional representations, or formal logic. One of the earliest and most fundamental discoveries of cognitive science was the heuristic nature of thought (Glaser & Chi, 1988; Newell, 1976); another was the fuzzy nature of such mental constructs as categories (Medin, 1989). So I do not see how these can be features that are critiqued in cognitive science, let alone insights that can be claimed for embodied cognition.

## Conceptual Metaphors

A core feature of both papers is their focus on identifying conceptual metaphors. Indeed, when I read work from the embodied cognition perspective, the identification of conceptual metaphors is where I learn the most. Uncovering tacit knowledge is never easy, but both of these papers—like much other work from the embodied cognition perspectives—display a genius at seeing the tacit structure in the speech of their participants.

A core question, from my point of view, is how conceptual metaphors relate to the sort of theoretical constructs identified in earlier work—constructs such as p-prims or my symbolic forms. Both sets of authors suggest that what they are describing is a new kind of resource, which supplements, rather than supplants, the list of resources identified by prior research. The primary conceptual metaphor identified by Brooks and Etkina is the caloric metaphor, which they describe as a 'linguistic resource'. Jeppsson et al. are also clear that what they are identifying is something additional, and is not intended to supplant resources described in earlier research. They say: 'We have suggested that conceptual metaphor may be added to the list of productive intuitive resources that contribute to science learning.'

Still, I come away with the sense that a great deal gets subsumed within the construct of conceptual metaphor, and that it unnecessarily breaks down boundaries that should be left up. Contrast, for example, the view laid out by Dedre Gentner across a wide range of articles. For Gentner, there have always been many types of 'domain comparisons' including 'abstraction' and 'analogy' (Gentner, 1983). In a later work, she opened up this taxonomy even more, describing a broad space of types of similarity relations (Gentner & Markman, 1997). Whether or not we accept Gentner's take on these issues, it is clear to me that not all domain comparisons are created equal. Let us consider one example from Brooks and Etkina. They talk about 'tunneling' as a metaphor that is used to describe how a quantum mechanical object can pass through a boundary. Are we certain that it is appropriate to describe this as a metaphor? There might instead be a common abstract structure (an abstraction) shared by many phenomena, and for which there is no particular source domain.

Perhaps the authors would object that what I am calling 'abstraction' is indeed covered in what they mean by 'conceptual metaphor'. My response is that this is precisely the problem: I am concerned that they are trying to make the notion of conceptual metaphor cover too many disparate reasoning phenomena.

In my own view, there is a diverse flora and fauna of entities of mind, and that we should be cautious about attempting to capture them all with a single theoretical construct that loses some of this diversity. To cite one example from the work of Jeppsson et al., the authors talk about conceptual metaphors associated with the levels of engagement of the speaker with the problem they are solving, for example, 'A Problem Solver Is an Owner/Observer of a System'. But I think we should ask ourselves whether these conceptual metaphors are really the same sort of beast as the other conceptual metaphors identified by the authors, such as 'Change of State Is Forced Movement'.

I have one last note about conceptual metaphors. I must admit that I do not understand how we are supposed to know when we have correctly identified a conceptual metaphor. How do we know that 'Change of State Is Forced Movement' is the right rendering in these instances? Part of the answer seems to be that the authors make specific bets about the level of abstraction and generality at which these conceptual metaphors live; there are theoretical assumptions that guide them. I also suspect that some sort of triangulation is required, but that it is not described in the papers because of limitations of space.

## The Body

At the start of this essay, I stated that I thought too much was being subsumed within the 'embodiment turn'. In particular, it seems that, according to Jeppsson et al., discussions of mental entities such as mental model, p-prims, and perhaps even symbolic forms are all part of the 'embodiment turn'. I want to officially raise my objection to being included as part of the embodiment turn (even though it would put me in very, very good company).

There are foundational questions at issue here, and there are limits to what I can say in this short essay. Researchers such as Lakoff claim that a mechanism such as conceptual metaphor is needed to solve the 'grounding problem' (Lakoff & Johnson, 1999). The idea is that, for us to truly understand something, it must be reducible to some common set of elements. It also helps greatly if these primitive elements are shared among humans. Though this view is not explicitly stated in the two papers, I believe it is at least tacit. At the very minimum, the authors have been strongly influenced by a research tradition built on this assumption.

In short, I do not see any need for, or any particular benefit in, any grounding of this sort. I do not see any problem with the notion that mental entities get their meaning from how they participate in a web of relations among themselves, and how they function in human action. In my view, there is no need for a special class of elements associated with the body. Note that this is in line with my more general view that there are a diverse flora and fauna of mind—a weaving together of diverse entities—with no single class that has a wide special importance.

## A Central Role for Language

I conclude with some comments on language. As I noted earlier, I believe that the attention to language is one of the important contributions in these papers, and one that distinguishes this work from much earlier work. Of course, the relationship between language and thought is fraught, and it is a relationship that has been wrestled with at least for decades, if not for centuries or millennia. Here, the authors are appropriately cautious, with particular attention paid by Brookes and Etkina. They say:

> With our data and methodology, it is not possible for us to make causal assertions about the impact of the caloric metaphor on students' reasoning, whether reasoning is being driven by language choices, or if students' underlying conceptions of heat are influencing their choice of language. What we observe from our data is that the caloric metaphor is connected to reasoning about heat that appears as state-function-like reasoning. (Brookes & Etkina, 2015)

The question to ask ourselves is whether Brookes and Etkina have made any progress in teasing apart this fraught relationship. Ultimately they say that the relationship must be 'bi-directional'. They also cite Jay Lemke as saying that 'concepts do not exist independently from their representations, and it is the representations themselves (graphs, equations, spoken/written language etc.) that constitute the concept itself' (Brookes & Etkina, 2015).

As I conclude this essay, I cannot resist adding my own thoughts on this issue that has consumed so many thinkers. I do not think we need to go as far as Lemke, and give up on a notion of concepts that sees them as located in the mind of individuals. Instead, my view is that we should see our knowledge and representational tools as having co-evolved, both phylogentically and ontogenically. This means that our knowledge is adapted by and for symbol use. Thus, we should expect to find many and multifarious ways in which knowledge and our representational tools will be finely tuned so as to work well with each other. We can expect no less. But, as far as sweeping claims go, we can say no more.

## Disclosure Statement

No potential conflict of interest was reported by the author.

## References

Brookes, D. T., & Etkina, E. (2015). The importance of language in students' reasoning about heat in thermodynamics. *International Journal of Science Education*. doi:10.1080/09500693.2015.1025246

Chi, M. T. H. (2005). Commonsense conceptions of emergent processes: Why some misconceptions are robust. *Journal of the Learning Sciences, 14*(2), 161–199.

Clement, J. (1994). Use of physical intuition and imagistic simulation in expert problem solving. In D. Tirosh (Ed.), *Implicit and explicit knowledge* (pp. 204–244). Hillsdale, NJ: Ablex.

diSessa, A. A. (1993). Toward an epistemology of physics. *Cognition and Instruction, 10*(2–3), 165–255.

Gentner, D. (1983). Structure-mapping: A theoretical framework for analogy. *Cognitive Science, 7*(2), 155–170.

Gentner, D., & Markman, A. B. (1997). Structure mapping in analogy and similarity. *American Psychologist, 52*(1), 45–56.

Glaser, R., & Chi, M. T. H. (1988). Overview. In M. T. H. Chi, R. Glaser, & M. J. Farr (Eds.), *The nature of expertise* (pp. xv–xxviii). Hillsdale, NJ: Erlbaum.

Jeppsson, F., Haglund, J., & Amin, T. G. (2015). Varying use of conceptual metaphors across levels of expertise in thermodynamics. *International Journal of Science Education*. doi:10.1080/09500693.2015.1025247

Lakoff, G., & Johnson, M. (1999). *Philosophy in the flesh: The embodied mind and its challenge to Western thought*. New York, NY: Basic Books.

Medin, D. (1989). Concepts and conceptual structure. *American Psychologist, 44*(12), 1469–1481.

Newell, A. (1976). Computer science as empirical inquiry: Symbols and search. *Communications of the ACM, 19*(3), 113–126.

Sherin, B. L. (2001). How students understand physics equations. *Cognition and Instruction, 19*(4), 479–541.

# Applying Conceptual Blending to Model Coordinated Use of Multiple Ontological Metaphors

Benjamin W. Dreyfus[†], Ayush Gupta[†] and Edward F. Redish
*Department of Physics, University of Maryland, College Park, MD, USA*

Energy is an abstract science concept, so the ways that we think and talk about energy rely heavily on ontological metaphors: metaphors for what kind of thing energy is. Two commonly used ontological metaphors for energy are *energy as a substance* and *energy as a vertical location*. Our previous work has demonstrated that students and experts can productively use both the substance and location ontologies for energy. In this paper, we use Fauconnier and Turner's conceptual blending framework to demonstrate that experts and novices can successfully blend the substance and location ontologies into a coherent mental model in order to reason about energy. Our data come from classroom recordings of a physics professor teaching a physics course for the life sciences, and from an interview with an undergraduate student in that course. We analyze these data using predicate analysis and gesture analysis, looking at verbal utterances, gestures, and the interaction between them. This analysis yields evidence that the speakers are blending the substance and location ontologies into a single blended mental space.

## Introduction

Energy is a central concept across the sciences in physics, biology, and chemistry. Understanding how students think about energy is essential to science education within each discipline and to making connections across disciplines. In our previous work (Dreyfus, Geller et al., 2014; Dreyfus, Sawtelle, Turpen, Gouvea, & Redish, 2014), we examined the conceptual demands placed by an interdisciplinary approach to energy in introductory physics, specifically when trying to understand the concept

---

[†]These authors contributed equally to the data analysis and production of this paper.

of negative energy. Negative energy becomes an important tool for understanding chemical bond energy (an important concept in introductory biology and chemistry) in terms of potential energy (an important concept in introductory physics). In many introductory physics courses, chemical energy is not an explicit topic of instruction, but becomes so when the course content is reformed with an emphasis on seeking interdisciplinary coherence between physics, biology, and chemistry (Dreyfus, Sawtelle et al., 2014; Redish et al., 2014). This poses instructional challenges.

Recently, many researchers who focus on developing learners' conceptual understanding of energy have emphasized the advantages of using a substance metaphor for energy. Conceptualizing *energy as a substance* is instructionally helpful because it tacitly brings along many useful properties in terms of accounting for the transfer and conservation of energy. However, it is difficult to extend the *energy as substance* metaphor to negative energy because we typically do not conceptualize substances as 'negative', or less than nothing. However, since energy has to be added to break a bond (chemical or nuclear), it is convenient to describe the potential energy of bound systems as negative (adding a positive quantity brings the energy of the system to 0). Dreyfus, Geller et al. (2014) have shown that drawing on the *energy as location* metaphor can be productive for developing a conceptual understanding of negative energy, and that experts as well as novices can productively coordinate the *energy as substance* and the *energy as location* metaphors.

While our earlier work shows that students and faculty can productively use both metaphors, in this paper we develop fine-grained cognitive models to demonstrate **how** experts and novices coordinate these two metaphors for energy. One of our central questions in this paper is whether the two metaphors remain separate but coordinated mental spaces, or are better modeled as a single mental space with the concept of energy taking on properties from both metaphors. In exploring this question, we draw on the conceptual blending framework proposed by Fauconnier and Turner (2002), and on gesture analysis (Goldin-Meadow, 2003). In addition to showing how ontological metaphors help understand how the concept of energy is built, this paper contributes to the existing literature in two ways: (i) illustrating that conceptual blending can be a useful tool for modeling concepts that cut across disciplinary boundaries, and (ii) illustrating that gestures can be a methodological tool in identifying the character of a blend.

In the following section, we briefly review research on instructional metaphors for energy and situate our research question within this work. Then we review Fauconnier and Turner's framework for conceptual blending and present how it can be applied specifically to generate cognitive models for energy metaphors. We then present our methods, followed by our analysis and results. We end with the implications for research and for instruction.

## Ontological Metaphors for Energy

Within physics education research, energy has long been a topic of interest (Beynon, 1990; Goldring & Osborne, 1994; Lawson & McDermott, 1987; Watts, 1983).

Recently, this interest has been renewed, with special emphasis on the metaphors that underlie experts' and novices' reasoning about energy (Amin, 2009; Brewe, 2011; Lancor, 2014; Scherr, Close, McKagan, & Vokos, 2012).

The research on metaphors for energy draws on the theory of conceptual metaphors developed by Lakoff and Johnson (1980/2003, 1999). In Lakoff and Johnson's framework, the way that we think and talk about abstract ideas is organized according to concepts grounded in physical experience. This structuring of one idea or concept in terms of another constitutes a metaphor, and the grounding of chains of metaphors in physical experience is referred to as *embodied cognition*.

Some metaphors are ontological, in the sense that they require mapping ideas across basic categories to which the ideas belong, for example 'viewing events, activities, emotions, ideas, etc., as entities and substances' (Lakoff & Johnson, 1980/2003, p. 25). An example (outside of physics) that Lakoff and Johnson present is *The Mind Is A Brittle Object*, represented in such sentences as 'Her ego is very *fragile*' and 'The experience *shattered* him' (Lakoff & Johnson, 1980/2003, p. 28). In a similar vein, *energy as substance* is an ontological metaphor that conceptualizes energy in terms of properties typically attributed to substances/matter, such as thinking of energy as being stored in some object, transferred from one object to another, and conserved (energy cannot be created or destroyed) (Brewe, 2011; Lancor, 2014; Scherr et al., 2012).

A second ontological metaphor for energy conceptualizes it as a vertical location (Amin, 2009; Scherr et al., 2012). In this metaphor, we think of 'more energy' as a higher location along a vertical axis (where the vertical axis represents energy values and might not correspond to actual physical positions[1]). In this paper, we focus on how experts and novices coordinate these two metaphors when talking about physical or biological phenomena.

This work on ontological metaphors takes place against the backdrop of an ongoing debate in the literature about the nature of learners' ontologies in physics. A prominent line of research by Chi, Slotta, and their colleagues (Chi, 2005; Chi & Slotta, 1993; Chi, Slotta, & de Leeuw, 1994; Slotta, Chi, & Joram, 1995) argues that physical concepts such as electric current, heat, and light belong to a 'scientifically correct' ontological category of emergent processes, that there is a cognitive barrier to flexibly transitioning between ontological categories for a concept, and that robust physics misconceptions about these concepts in novices result from miscategorizing these concepts in the incorrect ontology of substances or direct processes.

Recently, this framework has been challenged (Baily & Finkelstein, under review; Brewe, 2011; Gupta, Elby, & Conlin, 2014; Gupta, Hammer, & Redish, 2010; Hammer, Gupta, & Redish, 2011; Jeppsson, Haglund, Amin, & Strömdahl, 2013; Scherr, Close, & McKagan, 2012), arguing for a more dynamic view of ontological reasoning. This emerging line of argument demonstrates that experts as well as novices flexibly transition between the substance and process ontologies (including emergent processes) when reasoning about phenomena involving electric current, light, heat, energy, entropy, and quantum mechanics. Physics education researchers have argued for the productivity of conceptualizing non-material entities such as energy, gravity, and heat in terms of properties typically associated with substances (Brewe, 2011; Brookes, 2006; Gupta et al., 2014; Scherr et al., 2012). While we do not directly address this debate, our argument here touches on it in two ways: (i) our observations show how experts and novices can blend ontological metaphors with ease and this lends support to the idea that ontological cognition is flexible and (ii) we add gesture analysis to the methodology of investigating learners' and experts' ontological cognition.

## Conceptual Blending and Metaphors for Energy

In this section we show how metaphors for energy can be conveniently understood within the conceptual blending framework. This forms the background for our subsequent analysis of our video data.

Lakoff and Johnson's (1980/2003, 1999) conceptual metaphor theory proposes how metaphors arise when one domain (target) is cognitively structured in terms of the elements of another domain (source). The mapping allows aspects of the source domain to be transferred to the target domain. More recently, Fauconnier and Turner (2002) proposed a conceptual blending framework that can also be used to understand the cognitive processes that lie behind metaphors. Their framework shares many aspects with conceptual metaphor theory (Fauconnier & Lakoff, 2013).

In conceptual blending theory, input mental spaces are the cognitive constructs, instead of source and target domains. A blended mental space arises from mappings between entities in the two spaces, and the blended space can (and often does) include entities and relationships from one or both spaces. One key difference, however, is that conceptual metaphor theory mainly deals with entrenched metaphors (where the structuring of one domain in terms of another is relatively stable and long term) (Grady, Oakley, & Coulson, 1999). Blending theory, on the other hand, allows for local mappings between constructs—mappings that might not extend to entire conceptual domains and that might be relatively short-lived. Since we are not dealing with metaphors that are part of our everyday lexicon, or those that we can assume are particularly entrenched (especially in novice reasoning concerning energy), conceptual blending framework allows us greater flexibility in modeling the phenomenon of interest.

In physics education research, conceptual blending has been used to explain analogical scaffolding, the process by which students layer multiple levels of analogies to learn abstract ideas (Podolefsky & Finkelstein, 2007), and to model different ways that students reason about the propagation of wave pulses (Wittmann, 2010).

Understanding how and why conceptual blends occur requires dynamic principles—a model of the causes that drive blends. Fauconnier and Turner (2002) identify critical characteristics of the mental spaces (vital relations) and propose tendencies or forces involving these vital relations that drive blends (constitutive and governing principles). They note that there are many competing tendencies and the dynamical principles of blending theory identify these tendencies rather than predict unique results. Nevertheless, the proposed characteristics and principles (and subsequent elaborations and additions, e.g. Bache, 2005; Hougaard, 2005) provide tools for creating new descriptions of cognitive phenomena that can add useful insights.

*Vital relations* are relationships between input spaces that play an important role in conceptual blending. These include parameters of a situation such as time, space, cause–effect, part–whole, identity, and representation. One example that Fauconnier and Turner use is that of cold ashes in a fireplace. The way we interpret cold ashes in the fireplace as the remnants of (and resulting from) a past fire represents a conceptual blending: The input space of cold ashes is blended with the input space of fire in the fireplace through compression of various vital relations. Both the fire and the ashes occupy the same space but are separated in time (the fire was earlier and the ashes later). From the ashes we can infer that there were logs burning in the fire and that it is the logs that became ashes, and that it is the fire that caused that transformation. Thus, Fauconnier and Turner illustrate how the two spaces are connected by vital relations of time, space, change, and cause–effect (Fauconnier & Turner, 2002).

Principles identified by Fauconnier and Turner include compressing one vital relation into another, coming up with a story, pattern completion, and unpacking, among others. *Compression* is when we make connections across vital relations between two input spaces; for example, a graph that illustrates different historical recessions (using the same time axis for all of them) is compressing over time. But it is also possible to compress one vital relation into another (e.g. through representing time by space). *Pattern completion* is when we bring in additional elements to the blend by using existing patterns as additional inputs. *Unpacking* is the principle that the blend itself should include all the information necessary to reconstruct the entire network that produced the blend.

For the particular example we consider here, we do not need the full dynamic machinery of blending. The phenomenon we are reporting is well described as using the vital relations of space (the inputs might share locations, or not), part–whole (one input is a part of the other, e.g. when we see a picture of a person's face and identify the person, not just the face, we compress over this vital relation), and representation (one input is a representation of the other, e.g. a picture of an object and the object itself), driven by compression of vital relations, coming up with a story, and unpacking.

Each of the input spaces in a blend has a separate organizing frame. If only one of those frames is projected to organize the blend, then the conceptual integration network is *single-scope*; if the organizing structure of the blended space contains structure that comes from both input spaces, then it is *double-scope* (Fauconnier & Turner, 2002). Single-scope networks include conventional metaphors, in which one input (providing the organizing frame) is the source, and the other input (which is the focus of understanding) is the target. Double-scope networks are more complex, and Fauconnier and Turner have argued that double-scope blending is essential for the development of language and modern human cognition.

*Conceptual blending analysis of energy*

We describe blended ontologies for energy in terms of two levels of conceptual blending (Figure 1):

First, the *energy as a substance* and *energy as a vertical location* ontological metaphors are themselves described as blended spaces, formed by mapping between the energy input space and the substance and vertical location input spaces. Because these blended spaces take on the organizing structures of the substance and vertical location spaces, respectively, they are single-scope blends. If there is a structure to the energy concept in the absence of metaphors (e.g. understanding energy as a purely mathematical construct), this structure is not present in the metaphorical substance and location blends. Second, we hypothesize that the *energy as a substance* and *energy as a vertical location* ontologies become input spaces that form a blended substance–location ontological metaphor. Because this blend incorporates structures of both the substance and location spaces, it is a double-scope blend.

The process of conceptual blending involves selective projection from the input spaces: only some elements from each input space participate in the blend. The full concept of energy is complex and contains many elements; only a subset of these elements is projected into the *energy as a substance* blend, while a different subset is projected into the *energy as a vertical location* blend.

*Describing the energy as a substance metaphor as a conceptual blend*

To represent the *energy as a substance* metaphor as a conceptual blend, we first present what the mental space for 'substance' or 'things' might contain and what the mental space for energy might contain. Of course, these are partial representations of each idea, and the selection of items we represent in these mental spaces are already

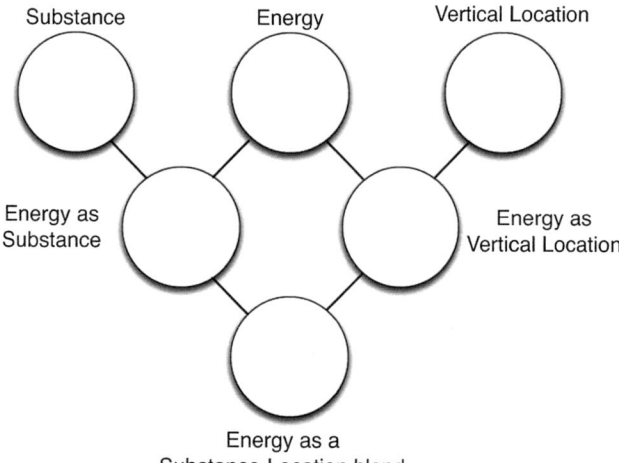

Figure 1.  The chain of mental spaces blended to form the concept of energy

guided by what we think would be relevant for presenting the metaphor as a conceptual blend. In that sense, this exercise is illustrative rather than explanatory.

From our everyday experiences, we can conceive of a mental space that incorporates how we think about substances: stuff moves from one place to another, can be stored, can be transformed (e.g. water can freeze into ice), and possesses object permanence (that is, objects cannot simply come into existence from nothing and cannot simply cease to exist).

The energy mental space contains many conceptual elements, which include the ideas that energy can be quantified; the sum total of the energy of a closed system is invariant; energy can be associated with a particular object (or a system of objects);[2] and there are different types of energy (kinetic, potential, etc.).

As illustrated in Figure 2, mapping elements from the energy space with elements from the stuff space can yield elements of the *energy as a substance* ontological metaphor as identified in Lancor (2014) and elsewhere (Brewe, 2011; Scherr et al., 2012).

The blended mental space is composed of object (or system of objects) and energy as entities that are now related through the organizational property of 'storage,' projected from the 'Stuff' mental space. Thus in the blended space: *energy is stored* in the associated object (or system of objects). This can also be conceptualized as a compression of the external vital relation between the ideas of association (energy space) and storage (stuff space) into an internal relation within the blended space. Also part of the blend is the mapping of invariance of total energy (of a system) over a process onto object permanence. In the blend, energy (like an object) cannot

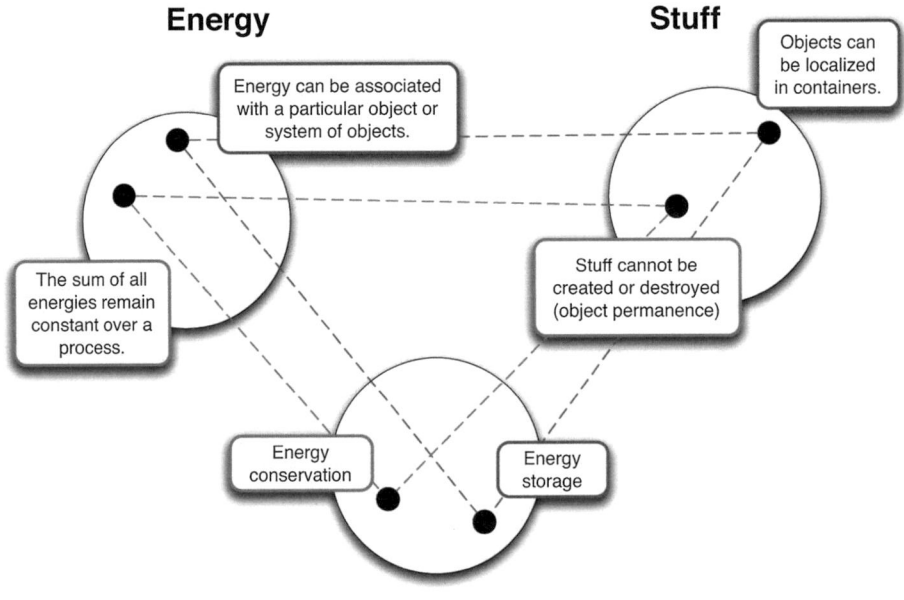

**Energy-as-substance**

Figure 2.   The energy as substance blend

be created or destroyed (without pre-existing in some form). In doing so, 'energy' and 'stuff' (the contained stuff) are compressed into a uniqueness relation: **Energy is stuff**.

Creating this blend implies that selective properties of the stuff or substances can potentially be projected onto energy. As we 'run' the blend, emergent meaning arises from elaboration: a decrease in associated energy of an object is seen as energy *leaving* that object (just as stuff can be taken out of a container, leading to a decrease in the amount of stuff in the container). *Energy conservation* would imply that we now need to account for the energy that leaves an object. The energy could be associated with another object (corresponding to putting stuff into a different container), whose energy thus increases (this is what is seen as energy moving from one object to another—*energy transfer*). Or, the energy is not seen as bound to another particular object (*emission of energy*). Similarly, just like adding stuff to a container, within the blend we can refer to *absorption of energy*. Finally, just as the conservation principle applied to stuff also means that stuff could be transformed from one state to another but not simply cease to exist, conservation of energy can also be understood within the blend through *transformation of energy* from one type to another (for example, the mathematical balance of kinetic and potential energy for an object falling under gravity can be understood in terms of a transformation from potential energy to kinetic energy).

The metaphor of *energy as substance* is so entrenched in our conceptual system, that most of the properties outlined above (emission, absorption, transfer, and transformation) are considered part of how we (experts and novices) talk about energy—properties we can see as having projected back onto the energy space. In what follows, we will use two of these emergent properties of the *energy as substance* space—energy can be absorbed and energy can be emitted—to form a subsequent blend with the *energy as location* space.

*Describing the energy as a vertical location metaphor as a conceptual blend*

The *energy as a vertical location* blend involves a different set of elements from the energy mental space: (i) energy is a state function (i.e. the energy of a system is only a property of the state of the system and is independent of the path that the system took to reach that state) and (ii) the energy of a system is a well-ordered quantity that can be more or less.

The relevant elements of the vertical location mental space are: (i) a vertical 1D coordinate is well ordered (and a coordinate can be conceptualized as being *higher or lower* than another coordinate) and (ii) a coordinate represents a physical location (objects are 'at' locations)

These elements are combined as in Figure 3. In the blended space, energy is mapped onto a vertical location coordinate through a uniqueness compression: energy *is* vertical coordinate. This is represented in the blend through the element 'More (less) energy is higher (lower).' Also, in the blended space, the state function of energy represents a physical location, giving meaning to expressions such as

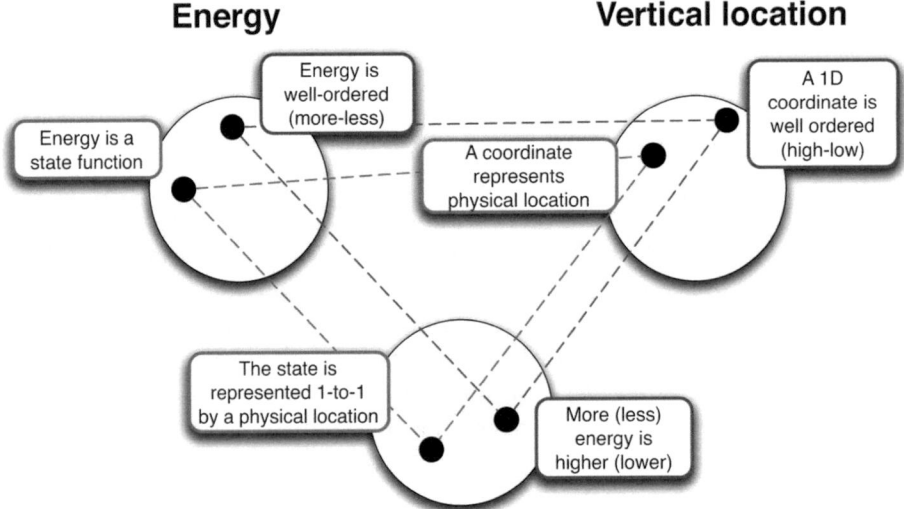

**Energy-as-vertical-location**

Figure 3.   The energy as vertical location blend

'Objects are **at** an energy level.' Within the blend, objects with more energy are 'higher' than objects with less energy.

As we 'run' the blend, new meaning emerges. We know that the negative gradient of the potential energy determines the force vector, which implies that an object (with no initial velocity) will move in the direction of the negative gradient of potential energy. We complete the blend by appending our physical experience with objects in space: that objects tend to fall downwards. Within the blend, new meaning arises, where we can think of the motion of objects within a potential energy gradient as 'tending to fall toward lowest potential energy'.

*Creating a new mental space through blending the energy as substance and energy as location metaphors*

We propose that the *energy as substance* and *energy as location* metaphors can themselves be blended to create a new blend that combines properties from both metaphor spaces. Here, we present what such a blended mental space would look like. Later, we present analysis of verbal data from experts' and novices' reasoning about energy to argue for the psychological existence of the blended mental space.

There are multiple ways that a blend between the substance and location ontologies for energy can manifest itself. We focus on one of these blends, shown in Figure 4, which we observe in our data. In this blend, the idea that objects absorb and release energy (from the *energy as substance* blended space) is mapped onto the idea that more and less energy corresponds to going up and down along a vertical

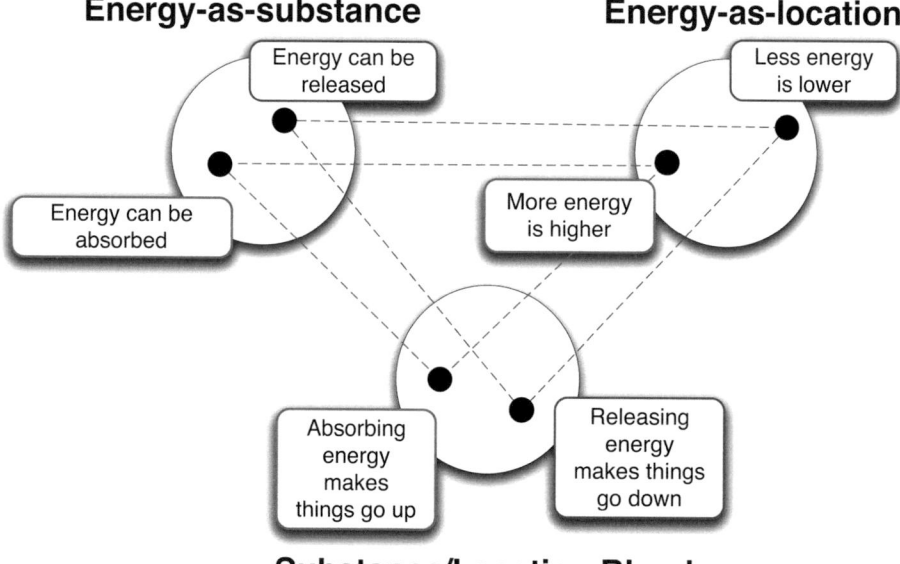

**Figure 4.** Blending the substance and location blends to describe emission and absorption of energy

energy dimension (from the *energy as vertical location* blended space). Note that absorption and release of energy were ideas we posited as emerging within the *energy as substance* blended space. These are ideas so entrenched in our everyday conceptualization of energy that we can think of them as reified elements within the *energy as substance* space.

In the new blended space, absorbing energy makes an object go up to a higher energy-location, and releasing energy makes it go down to a lower energy-location. Thus in the blended space, movement along the vertical energy dimension comes to be associated with processes of energy transfer and energy transformation. In this blended space, energy functions simultaneously as both a substance (being moved and transferred) and a vertical location.

Additionally, new meaning arises through the blending process. Absorption or emission of energy are not just associated with higher or lower energy location, but are seen as processes that cause an object to move up or down the energy axis. This sense of causality emerges within the blend through pattern completion (Fauconnier & Turner, 2003). A more explicit example of this sense of caused motion is apparent

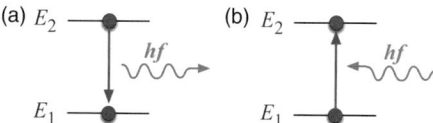

**Figure 5.** Representation of absorption and emission of energy from a quantum system (Wikipedia, 2014)

in how we explain electron transitions. Figure 5 (Wikipedia, 2014) shows a photon carrying energy out of or into a quantum system by combining the representation of energy in the system as higher and lower levels (location) and the addition or removal of energy as adding or taking away a photon (substance). The electron's movement from one energy level to another is conceptualized as being precipitated by the absorption or emission of the photon.

So far, we have proposed a structure for the conceptual blending of the substance and location ontological metaphors for energy, and now we seek to illustrate what such blended conception might look like when experts or novices explain energy phenomenon. In our analysis of data we draw heavily on gesture analysis, which has previously been a useful tool in science education. In the next section, we briefly discuss the research on gestures and their use in science education.

## Gestures as Tools for Cognitive Analysis

Gestures are spontaneous hand movements of speakers that are tied to speech and are in the service of communication (Goldin-Meadow, 1999, 2003; Scherr, 2008). Gestures can be classified into multiple categories (McNeill, 1992), including deictic (pointing), iconic (emulating concrete objects), and metaphoric (visually representing abstract concepts).

In science education research, gestures have been analyzed as essential elements of how students communicate and reason about scientific concepts. Gestures play an important role in the development of scientific language and can precede verbal communication (Roth, 2000; Roth & Lawless, 2002; Roth & Welzel, 2001). Gestures can be a source of evidence of the content of students' ideas, particularly for ideas that students are constructing in the moment (Scherr, 2008). Examples include hand motions illustrating the trajectory of a projectile (Scherr, 2008) or the directions of forces and velocities (Chase & Wittmann, 2013).

Gestures can often embody conceptions that are not verbalized in speech. Within conceptual metaphor theory and in conceptual blending theory, metaphors structure not just speech but also the underlying cognition; this implies that non-verbal channels of communicating thinking, such as gestures, could provide additional ways to investigate metaphorical reasoning (Cienki & Müller, 2008). Gestures have been used as part of conceptual blending analysis in previous work, but with a different focus. Edwards (2009), in the context of mathematical reasoning about fractions, used the gestures themselves as one of the input spaces to the blend, and showed how the blended space integrated conceptual and gestural elements. Wittmann (2010) uses gestures as one channel of communication that is analyzed for evidence of conceptual blending. He presents a case study in which students blended one input space of an observed wave pulse on a string with another input space of a thrown ball (leading to the incorrect conclusion that creating a wave on a string with a faster flick of the wrist would cause the wave to move faster). Students' gestures (for example, moving the hand in a motion to produce a wave pulse on a string) served as evidence for this conceptual blending. Our investigation is different from Edwards' (but similar to Wittmann's) in that we are using gestures as evidence for an underlying conceptual blend rather than considering the gestures themselves as an input to the blend.

In the methodology section we describe the types of gestures we specifically draw on to support our analysis of conceptual blending.

## Methodology

### Data Collection and Selection

For this paper, we selected data excerpts from an introductory physics course for undergraduate life sciences majors (Redish et al., 2014). This course is unusual in that it emphasizes chemical energy as a way to build interdisciplinary coherence across physics, biology, and chemistry (Dreyfus, Gouvea et al., 2014), thereby providing us with opportunities to see how students responded to the concept of energy across disciplinary contexts. We collected video recordings of every class for the first two years. We also conducted semi-structured interviews with students in the course on a variety of topics.

The data set was not collected mainly for the purposes this paper is addressing, but the research questions driving the data were loosely organized around the notion of experts' and novices' ontologies, and hence these data were suitable for mining to address the question we pose here.

The first selection from our data set happened in the course of analyzing the data for purposes outlined elsewhere (Dreyfus, Geller et al., 2014). We viewed video data from classroom and interviews, roughly tagging places where participants tended to reason about energy using substance-based and location-based ideas. This tagging was often done in real time, drawing on verbal phrases such as 'you're at this energy' or 'put energy into the [object]'. For the purpose of the argument presented in this paper, we revisited the tagged episodes and selected out a subset of episodes where both substance- and location-based ideas for energy were in use. These episodes were then analyzed in greater detail as outlined below.

We chose two episodes for analysis:

- One episode from classroom video from Spring 2013. It depicts the physics professor, 'Dr Farnsworth' (pseudonym), reasoning about energy during lecture in service of explaining chemical bond energy to students. The episode caught our attention because of Dr Farnsworth's use of both substance-based and location-based phrases seamlessly within a short segment of speech.
- A second episode from an interview with a student in the course, 'Betsy' (pseudonym), conducted during Spring 2013. Betsy's interview focused on explaining energy transfer and energy transformation in the context of adenosine triphosphate (ATP) metabolism and other biological processes.

The episodes selected for analysis are intended to be illustrative of what ontological blending looks like. We are not attempting to make empirical claims about the prevalence of blending (relative to using a single ontology, or shifting discretely between two separate ontologies), and therefore we are not concerned with questions of how representative or typical these data are.

*Analysis Methodology*

We identify ontological metaphors in our data by looking at two different channels: words and gestures. To analyze the verbal data, we modify slightly the method of Slotta et al. (1995) to fit our particular needs. For gesture analysis, we draw on Goldin-Meadow (2003).

*Predicate analysis.*    To analyze the verbal data, we use a version of predicate analysis. Slotta et al. (1995) define predicates as 'words, phrases, or ideas whose presence in a spoken explanation is taken to reflect an underlying ontological attribute'. Within Chi's and Slotta's framework, attributes belong to ontological categories. The top-level categories that Chi and Slotta pose are matter/substance and process, and these are posited to be mutually exclusive (Chi, 2005; Chi et al., 1994). The predicates that we use to identify instances of the substance ontology for energy are essentially the same as those used by Slotta et al. (1995). Examples found there include *move* ('energy goes', 'energy flows' are example utterances for coding for the *move* attribute in relation to energy), *contain* ('stores energy', 'put energy in'), and *quantify* ('a lot of energy', 'some of the energy'). A small number of process predicates are found in our data as well, but are not a central part of our analysis. In addition, the other ontological category of relevance to our analysis is location, and predicates that act as markers for location. Location predicates are not included in Slotta et al.'s analysis, so we clarify here what is and is not included in this category.

Our analysis is about ontologies for energy, and so we only code a word or phrase as a location predicate if **energy** is described as a location, and some physical object or system is described as being at (or going to or from) that location. Examples include 'here', 'go', 'up', and 'down', when those refer to energy as location. However, 'energy goes there' is coded as a **substance** predicate for energy, because the energy is described as being **at** a location (which is an attribute of a substance), not as **being** a location. A location metaphor requires that the location stand for the quantification of the energy, not its spatial location. As a result, the predicates have to be evaluated in context.

*Gesture analysis.*    As with predicates, we posit that gestures can help in providing information on the ontological categories the speaker is drawing on at a given moment. To attend to gestures, we did our video analysis in layers. Analyzing the video jointly, in the first pass, we attended to the content of the speech, trying to make sense of what the speaker was saying. Then we made another careful pass closely attending to the gestures that co-occurred with speech. In this next pass, we would often go through the video frame by frame trying to note the gestures. Here, we posited multiple interpretation of what a particular gesture could be depicting or indexing. In order to understand what meaning participants might be making in a particular moment, we layered the gestures onto the verbal utterance to select which particular interpretations of gestures and speech provided the greatest explanatory power and coherence of meaning.

In our analysis, two kinds of gestures become meaningful: (i) metaphorical and iconic gestures that represent objects and ideas being talked about and/or aspects of the processes being described in speech and (ii) indexical gestures that point to something in the real material setting or in the representational space set up by the speaker at a particular moment. For example, pointing to a location on a physical representation of the vertical energy axis would support the idea that the speaker is thinking of *energy as vertical location*. On the other hand, a gesture tracing the trajectory of energy out of an object (being released into the surrounding air or being transferred to another object) will be taken as indicative of the *energy as substance* metaphor.

Our analysis of gestures is interpretive rather than following a systematic coding scheme. As such, it is difficult for us to make an exhaustive list of what particular gestures are coded for which ontology or metaphor. In lieu of that, we provide enough details in the data analysis so that the readers can draw their own conclusions and evaluate our gesture analysis.

We note that in some of the previous gesture literature (mostly work with children), mismatches of gesture and speech are taken to indicate a destabilization of the child's mental state or a transitional mental state and even a 'readiness to learn'. A standard example is children performing conservation tasks (water poured from a narrow into a wider glass) using gestures that contradict their words (e.g. showing 'wider' when they said 'taller'). (Church & Goldin-Meadow, 1986; Scherr, 2008). However, in cases where speech is metaphorical, gesture–speech mismatches might signal something else: that speech is communicating ideas from one mental space (target space, often) while gestures are communicating ideas from a different mental space (source space, often) (Forceville & Urios-Aparisi, 2009). Our interpretation of speech–gesture mismatch here aligns more with the latter interpretation. In our first episode, we are looking at a physics expert using gestures and language to describe an abstract complex concept that is not easily described in physically direct terms. In such situations, experts often use multiple ontologies in their professional interactions (Gupta et al., 2010). Gesture–speech mismatches can indicate that a speaker is thinking about two concepts at once (Chase & Wittmann, 2013), in a way that is more reliable than looking at speech alone, since speech is sequential, while gestures can be another simultaneous channel of communication. We take the gesture–speech mismatches in this case to be evidence of ontological blending, as explained below, and we support our argument with specific examples.

*Inferring blended ontologies.* We use predicates and gestures to infer whether participants are drawing on *energy as substance* and *energy as location* metaphors. To infer that participants are drawing on a blended space of these two metaphors we rely on

(i) coordination of meaning, (ii) temporal proximity, and (iii) speech–gesture mismatch. When an element from one metaphor space is mapped onto an element from the other metaphor space in the verbal utterance (an identity mapping within the conceptual blending framework) we take that as evidence of a form of compression of an external relation between elements in two different mental spaces into an internal relation within the blended space. Temporal proximity also plays a strong role in this inference. When ideas from two different metaphor spaces are coordinated within the span of a sentence or a phrase, this lends greater evidence that these ideas are not merely coordinated across mental spaces but might be compressed into a single blend. Finally, we note instances in the data where the content of verbal utterance draws from one metaphor space while the co-occurring gestures draw from another metaphor space in a way that meaning-making of the utterance requires us to coordinate the mismatched speech and gesture. Such instances indicate to us that both metaphors are being drawn on simultaneously to express a single integrated idea. Again, such temporal proximity (sometimes less than the span of a second), while not proof that ideas are blended rather than coordinated across the two metaphors, raises the plausibility of the existence of the blended mental space. Within Fauconnier and Turner's framework too, most examples of blended space tacitly rely on such mappings between mental spaces within a short time span.

## Data and Analysis

In this section, we present and analyze transcript segments from the data. The initial predicate analysis has been indicated in the transcripts themselves: substance predicates for energy are <u>underlined,</u> and location predicates for energy are indicated in **bold**. Gestures are also described *(in parentheses and italics)* in the transcript, along with selected still frames shown in the figures.

*Episode 1: The Substance–Location Blend in Expert Physics Reasoning*

This segment from Dr Farnsworth comes from the introductory physics for life sciences (IPLS) course. Dr Farnsworth was going over a quiz in class, and immediately before this segment, he drew a graph of a Lennard-Jones interatomic potential[3] on the chalkboard (Figure 6(a)), a representation that had been used frequently in the course for energy associated with chemical bonds. During this segment, Dr Farnsworth draws additional lines on that graph, shown in Figure 6(b).
    Prof. Farnsworth starts with:

> Prof. Farnsworth: If the two atoms are apart *(holds two hands apart, see Figure 7)* and form a bond *(moves hands together)*.

At the start of the quotation, Prof. Farnsworth held his hands in an iconic gesture as if holding the two atoms in his hands, the distance between his hands depicting the 'apart'-ness of the atoms (Figure 7). In the next portion of that sentence, as he

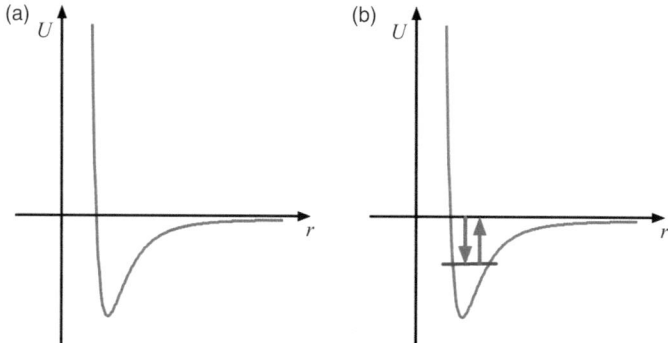

Figure 6.   (a) Lennard-Jones potential and (b) Lennard-Jones potential with the inscriptional marks (horizontal line and vertical arrows) made by Prof. Farnsworth

said, 'and they form a bond', he brought his hands closer. While the gesture depicts the two previously indicated atoms coming closer, the speech only refers to 'forming a bond' with no reference to distance. It is only through such speech–gesture coordination that we can make sense of the notion that, as the atoms form a bond, their physical distance is reduced. In the next portion of the utterance, Prof. Farnsworth continues with what this means with respect to energy:

> Prof. Farnsworth: **they drop down** (draws a horizontal line about halfway down the well, Figure 8(a)) **to here** (draws a vertical arrow from the top of the well to that line, see Figure 8(b)) and release that much energy (see Figure 8(c)).

As Prof. Farnsworth finished the earlier utterance, he turned to the board. This movement changes the operational space in which he can gesture. Previously, the motion of his hands in the three-dimensional space in front of him was depicting motion of atoms in physical space; the blackboard, on the other hand, has inscribed on it the Lennard-Jones potential as shown in Figure 6(a) in which energy is typically represented on the vertical axis, while physical distance between atoms is represented on the horizontal axis. As we can see from Figure 8(a), the inscription on the board does not mark out the vertical and horizontal axes, but within the context of previous discussions in the course, we can take this as accepted interpretation of those axes.

Figure 7.   (a) Prof. Farnsworth holds two hands apart, the shape of the hands representing two distinct objects (in this case, atoms). (b) Gesture depicts the atoms coming closer

Figure 8.  Prof. Farnsworth draws (a) a horizontal line about halfway down the well, (b) a double-tipped vertical arrow from the top of the well to that line, and (c) a vertical arrow from the horizontal line in the middle of the well to the top of the well. (Note that the yellow arrows added to the figure represent his hand motion as he produces the inscriptions on the board)

Simultaneous to saying, 'and drop down to here', Prof. Farnsworth draws a horizontal line halfway between the horizontal axis and the bottom point of the Lennard-Jones potential and then draws an arrow pointing downwards from the horizontal axis to the drawn horizontal line. By itself, the utterance ('and drop down to here') could have been a reference to the spatial location of the atoms changing as they form a bond, with atoms dropping in physical space. So, relying only on the verbal utterance, it is unclear whether we can code 'drop down' and 'to here' as reflecting predicates for the energy as vertical location metaphor. But when we attend to the co-occurring gesture, we can tie the 'drop' to the downward pointing arrow along the vertical energy dimension and 'here' as a location on the vertical energy dimension (and not the physical space), leading to the meaning that the atoms, when bonded, are not at a new physical location (in three-dimensional space) but at a new (metaphorical) energy-location. 'Drop down' and 'to here' can now be coded as predicates for the *energy as vertical location* metaphor. Thus, in this part of the utterance we argue that the fine-grained coordination of speech and gesture are producing the *energy as vertical location* metaphor.

In the next moment, Prof. Farnsworth's speech refers to the amount of energy being released as the atoms form the bond. The word, 'release' would refer to the substance attribute 'supply' per Slotta et al. (1995), suggesting that the speech in this moment is drawing on the *energy as substance* metaphor. However, gestural evidence adds further meaning to this moment: As Prof. Farnsworth utters 'released', he draws an arrowhead on the top of the earlier downward pointing arrow, suggesting that the amount of energy released is marked by the difference along the vertical energy dimension. In this coordinated production of speech and gesture, within the span of a second in which it is executed, the speech refers to the *energy as substance* metaphor, while the gestures draw on the *energy as vertical location* metaphor. Additionally, in representing not just energy as vertical coordinate, but also providing a visual sense of the amount of energy released ('that much' represented by the arrow) the inscription on the board becomes a single representational space that binds the two metaphors. In this part of the utterance, we argue that the fine-grained coordination of speech and gesture are producing the substance–location blend for energy.

There is yet another way in which this first sentence of the utterance enacts the substance–location blend. The two phrases, 'and drop down to here' and 'and release that much energy' are not disjoint phrases that just happen to be in temporal proximity in the utterance, but are conceptually bound so that the full physics implications of the sentence can be understood only by taking the phrases together. The motion of the atoms from one energy-location to another (a change in the energy state function of the system) is causally linked with the release of energy. This kind of causal elaboration is an emergent property of the substance–location blend as described before. This causal link becomes more explicit in the following segment:

> Prof. Farnsworth: And because **that's where they are** *(holds hands with the palms down, imitating the shape of the potential well)*, **at that negative energy**, that's equal to the <u>energy you have to put in</u> to get them back apart *(gestures getting the atoms apart)*. So it's just about *(puts one hand above the other, and moves hands up and down, out of phase)* **where you're going** *(turns to chalkboard)*, that when you form a bond *(puts hand horizontally at top of well)*, you're **dropping down** *(moves hand downward to horizontal line in middle of well)*, and if you **come in at this energy** *(redraws horizontal line at top of well)* you <u>gotta get rid of this much</u> *(draws another vertical arrow from the top of the well to the horizontal line in the middle)*. But if you're **down** *(holds hand horizontally at top of well)* **here** *(moves hand to horizontal line in middle of well)* and want to **get back up to here** *(moves hand back to top of well)*, you <u>gotta put in this much</u> *(draws vertical arrow from horizontal line in middle of well to top of well, see Figure 9)*.

Note that the phrasing 'if you ... get back up here, you gotta put in that much [energy]' indicates a causal dependence between energy transfer and motion of atoms along the vertical axis of energy.

In this segment, the pattern of verbal utterance and gestures that we detailed above is sustained: (i) substance and location predicates are interlaced within the same sentence, (ii) releasing (or putting in) energy is mapped onto moving down (or up) in energy-location, and (iii) the physical inscription of the Lennard-Jones potential is maintained as a space that represents energy as vertical location with arrows and distances along that axis representing amounts of energy released or put in. This

Figure 9. Prof. Farnsworth draws upward arrow to indicate the amount of energy needed to break the bond

indicates that Prof. Farnsworth is not just switching back and forth between the two metaphors, but that ideas from the two different mental spaces are blended into a single mental space. In this blended space, adding (or removing) energy to the system of two atoms does not just increase their energy; it moves them to a higher (or lower) energy-location. The contractions from individual mental spaces to the blended space are reflected in the speech–gesture 'mismatch', where the predicate from one ontological category is accompanied by gestures from another.

### Episode 2: The Substance–Location Blend in Novice Physics Reasoning

Betsy was an undergraduate pre-medical student enrolled in the IPLS course taught by Prof. Farnsworth. This interview was the second of two interviews with her that focused on reasoning about energy. The interviewer drew two Lennard-Jones potential graphs on the whiteboard (Figure 10), told Betsy that they represent two different systems, and asked Betsy which graph represented more energy. The question was intentionally underspecified: on the one hand, the minimum of the graph on the right (representing a bound state) is at a higher energy than the minimum of the graph on the left. On the other hand, the difference in energy between the bound state and the unbound state (shown as zero on the graph) is greater for the graph on the left. Betsy saw through the trick question and explained why either graph could be the answer, depending on the interpretation of the question.

The interviewer next followed up on an earlier conversation about ATP hydrolysis, and asked:

| | |
|---|---|
| Interviewer: | So between these two graphs, if one of these represented ATP and one of them represented ADP which one would you say was which? |
| Betsy: | (immediately writes ATP on the right and 'ADP+P in water' on the left) |
| Interviewer: | Okay, why? |
| Betsy: | Because in biology they always assume that it's in water 'cause the whole system is mostly made up of water. So if I put these two graphs together *(draws new graph with two valleys, see Figure 11)*, so this is ATP *(labels shallower well as 'ATP')* and it takes a little bit of energy to put in to get ADP *(traces up along graph from ATP well and then back down into deeper well, and labels deeper well as 'ADP', see Figure 12)*, but ADP is much more stable than this *(points to ATP)*, and this is because the phosphate reacts with the water and forms a really stable. So it's **in a well** but it **falls into a deeper well** once the bond breaks. I'm pretty sure. |

Here, Betsy's co-occurring gestures and speech draw on resources from different ontological metaphors. As she says, 'it takes a little bit of energy to put in' (reflecting the quantity and absorb attributes, per Slotta et al., 1995), she traces a path upward along the graph to indicate what happens to the system when energy is 'put in'. The vertical location on the graph represents energy. So we interpret Betsy's actions here to mean that, as energy is put in, the system moves up in energy-location. This location ontology is reinforced with Betsy's subsequent utterance that the system is 'in a well' and 'falls into a deeper well' (the well is not a spatial location, but a location along the

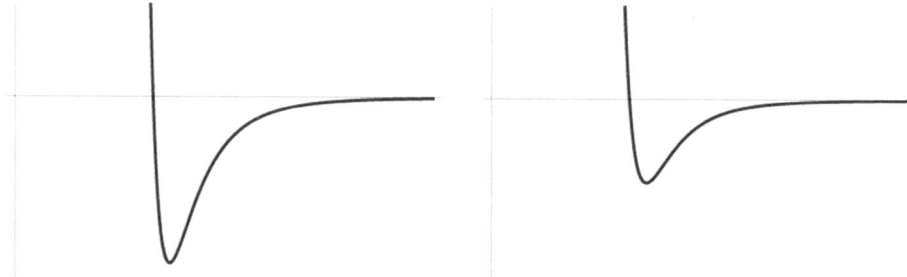

Figure 10.   Lennard-Jones potential graphs for two interacting atomic systems, as drawn in the Betsy interview. The vertical axis represents energy, and the horizontal axis represents the distance between atoms

energy dimension). The coordinated speech and gestures drawing simultaneously on 'energy can be absorbed or put into an object' and 'objects are at energy locations' reflect the blended element 'absorbing energy makes things go up'.

A few turns later in the conversation, Betsy says, 'They call ATP a high-energy bond', and the interviewer follows up on this:

Interviewer:    Yeah, why do they—what does that mean when they call it a high energy bond?
Betsy:          Yeah, it doesn't actually have—Professor [Farnsworth] was talking about that, it doesn't actually <u>hold energy</u> *(gestures as if 'holding' something, see Figure 13(a))*, like it's not—like, the bond itself doesn't <u>have a lot of energy</u>, but it's the fact of breaking it and forming this *(points to ADP)* is even more—is **even lower energy** *(gestures up and down, see Figure 13(b))*. So you can get—the difference between **here** and **here** *(labels vertical difference between two wells, see Figure 13(c))*, this is <u>the energy that's given off</u> to drive the rest of the system. Or drive the rest of the reaction.

Here, again, we see Betsy interleaving substance and location ideas (reflected in her verbal utterances and gestures). 'Hold energy', 'have a lot of energy', and 'energy that's given off' indicate contain, quantify, and supply substance-attributes (per Slotta et al., 1995). The utterance of 'even lower energy' is coordinated with gestures

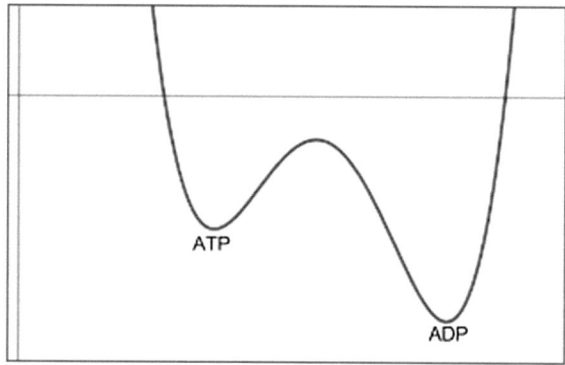

Figure 11.   The energy diagram that Betsy draws, representing ATP and adenosine diphosphate (ADP) on the same graph

Figure 12.   Betsy traces up along graph from ATP well and then back down into deeper well
(labeled as ADP)

to indicate lower energy-location. And the difference in positions of the energy locations ('the difference between here and here') is referred to as the amount of energy that is released. We take this close integration of the two metaphors as evidence of blending the two mental spaces to produce the idea (in blended space) that 'releasing energy makes things go down'.

This blending enables representing energy conservation in the location space. While conservation was already an affordance of the substance space (putting in

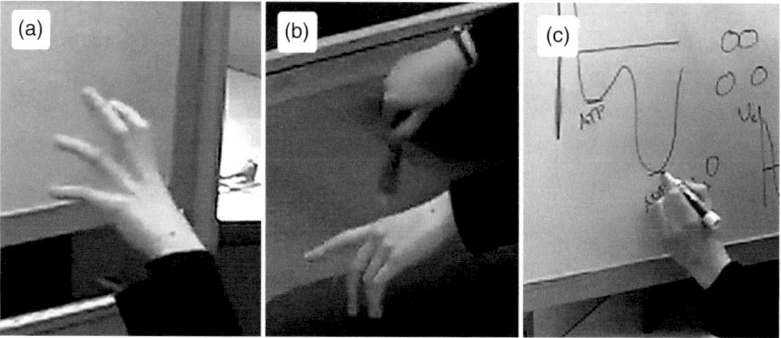

Figure 13.    (a) Betsy gesturing for 'doesn't actually hold energy'; (b) Betsy gesturing for 'is even lower energy'; and (c) Betsy's diagram for 'between here and here'

stuff from elsewhere increases the total amount of stuff inside, and losing stuff to the outside decreases the total amount of stuff on the inside), conservation is not represented in the location space on its own (going to a higher or lower location is not obviously associated with changes in the surrounding world). However, in the blended space, changes in location ('the difference between here and here') are associated with the transfer of a substance ('the energy that's given off') and therefore with interactions with the surroundings.

## Conclusions and Implications

We present illustrative cases on how a physics professor and an undergraduate student in an interdisciplinary introductory physics course for life sciences majors draw on two different metaphors for energy. We argue that for both of these subjects, the two metaphors are integrated into a single blended mental space as evidence through our analysis of their overlapping speech and gestures.

Our results offer further support for the value of viewing ontologies in physics not simply as fixed, but as a dynamic tool for building powerful conceptual models of abstract and difficult concepts (Gupta, Elby et al., 2014; Gupta, Hammer et al., 2010; Hammer et al., 2011). However, our analysis here does not directly address the argument for ontological commitments leading to misconceptions, as presented by Chi and Slotta (Chi, 2005; Slotta & Chi, 2006), since their argument pertained to the categories of substance and emergent processes, rather than that of location.

Through this paper we also contribute to the growing trend of multi-modal analysis (Stivers & Sidnell, 2005), attending to not just verbal utterances of learners but also other embodied modalities, to create models of learners' cognition. We illustrate one possible way in which speech and gesture can be co-analyzed to make sense of learners' and experts' cognition. In particular, we show how Fauconnier and Turner's conceptual blending theory in conjunction with gestural analysis can provide useful machinery for understanding how novices and experts blend multiple metaphors when reasoning about science concepts such as energy.

As has been well documented, expert knowledge (Chi, Feltovich, & Glaser, 1981; Machamer, Darden, & Carver, 2000; Reif, 1995) is not just about learning sets of independent facts; it is about organizing and coordinating facts and principles, seeking local and global coherence, and being able to generate sensible stories about complex phenomena. If we are to be able to understand students' progress along this dimension, we need analytic tools that can help us both understand expert knowledge structures and make sense of how students acquire and create such structures. While a knowledge-in-pieces theoretical frame (diSessa, 1993; Hammer, 1996; Smith, diSessa, & Roschelle, 1994) can help us make sense of the resources students can activate when faced with a learning task, we need more analytic tools to see how these resources interact (Gupta & Elby, 2011) and generate new knowledge structures. We believe that cognitive blending offers one such tool (see also Wittmann, 2010). The sort of analysis we have done here, using the machinery of conceptual blending to model the dynamics of a learner's cognition can potentially help us give us insight into how students learn to build new knowledge.

In particular, many science concepts such as energy are understood primarily through metaphors, often through multiple metaphors. Expertise in physics is not constituted in learning a single canonical way of reasoning about a concept or onto-logically categorizing that concept; rather, it is constituted in coordinating multiple metaphors and ontological categories to flexibly understand and apply that concept in different contexts. We expect that the kind of analysis done here can not only provide a better understanding of what students are doing in their path toward devel-oping expertise, but also offer guidance in how to create learning environments that facilitate that development. Our previous work provides an example of this. The context was the NEXUS/Physics course described above and the instructional goal was the extension of the traditional treatment of energy in introductory physics to include a discussion of chemical energy and bonding (Dreyfus, Gouvea et al. 2014). Our ontological research with this population (Dreyfus, Geller et al. 2014) and our instructional experience shows that many students apply the 'energy as sub-stance' metaphor and have difficulty making sense of negative potential energy in general and binding energy in particular. This led us to emphasize the fact that the zero of potential energy is arbitrary and in some cases, convenience leads to a choice that makes potential energy and even total energy negative. These lessons provide a scaffold for students to build the metaphor of 'energy as location'. This is done at the beginning of the discussion of energy in traditional contexts so it is not a new idea when the topic comes up in the context of chemical bonding. A critical element of this instruction is the use of visual representations that allow a geometric interpretation of the negatives (potential energy graphs with negative values and energy bar charts with bars that can be positive or negative).

For the current results, we plan to extend our previously developed examples and add new ones to focus on adding and removing energy. This will mean bringing in visual representations that blend the metaphors (such as Figures 5 and 6). The exposure to the blended energy metaphor will prepare students for later work on spectroscopy in chemistry where such diagrams are used extensively and quickly become quite complex.

This illustrates how a careful theoretical analysis of cognitive models can work with instructional development to better integrate and reconcile distinct mental models, as often occurs, for example, in articulating instruction among different disciplines (Dreyfus, Sawtelle et al., 2014).

## Acknowledgements

The authors thank the rest of the NEXUS/Physics research team (Ben Geller, Julia Gouvea, Vashti Sawtelle, and Chandra Turpen), the UMD Physics Education Research Group, Michael Wittmann, and two anonymous reviewers for discussions and feedback.

## Disclosure statement

No potential conflict of interest was reported by the authors.

## Funding

This work was supported by the National Science Foundation under the Graduate Research Fellowship (DGE 07-50616) and TUES DUE 11-22818, and by the Howard Hughes Medical Institute under the NEXUS grant.

## Notes

1. Typically, the energy axis does not represent location, but in the case of gravitational potential energy in the flat earth approximation, the potential energy is directly proportional to location and the graph of the height looks just like the graph of energy. This example provides a useful pedagogical introduction to this blend.
2. Note that the idea 'energy is associated with a particular object' is not a required property of energy, just a conceptual idea that is used in some circumstances. For example, while potential energy may be described as 'belonging to a particular object' when one member of the interacting pair is much larger than the other (the potential energy of a thrown ball, or the potential energy of an electron in an atom), in other circumstances energy is associated with multiple objects rather than a single one (the relative kinetic energy of two atoms or the potential energy of an electron–positron pair).
3. The Lennard-Jones potential is a commonly used model of the interaction potential energy between two atoms (Jones, 1924).

## References

Amin, T. G. (2009). Conceptual metaphor meets conceptual change. *Human Development, 52*(3), 165–197.

Bache, C. (2005). Constraining conceptual integration theory: Levels of blending and disintegration. *Journal of Pragmatics, 37*(10), 1615–1635.

Baily, C., & Finkelstein, N. D. (under review). Ontological flexibility and the learning of quantum mechanics. Submitted to *Physical Review Special Topics—Physics Education Research*. Retrieved from arXiv:1409.8499 [physics.ed-ph]

Beynon, J. (1990). Some myths surrounding energy. *Physics Education, 25*(6), 314–316.

Brewe, E. (2011). Energy as a substancelike quantity that flows: Theoretical considerations and pedagogical consequences. *Physical Review Special Topics—Physics Education Research, 7*(2), 020106 (14 pp.).

Brookes, D. T. (2006). *The role of language in learning physics* (PhD dissertation). Rutgers, The State University of New Jersey, New Brunswick. Retrieved from http://adsabs.harvard.edu/abs/ 2006PhDT . . . . . . .238B

Chase, E. A., & Wittmann, M. C. (2013). Evidence of embodied cognition via speech and gesture complementarity. In N. S. Rebello, P. V. Engelhardt, & C. Singh (Eds.), *AIP conference proceedings*, Philadelphia (Vol. 1513, pp. 94–97). American Institute of Physics. Retrieved from http:// dx.doi.org/10.1063/1.4789660.

Chi, M. T. H. (2005). Commonsense conceptions of emergent processes: Why some misconceptions are robust. *Journal of the Learning Sciences, 14*(2), 161–199.

Chi, M. T., Feltovich, P. J., & Glaser, R. (1981). Categorization and representation of physics problems by experts and novices. *Cognitive Science, 5*(2), 121–152.

Chi, M. T. H., & Slotta, J. D. (1993). The ontological coherence of intuitive physics. *Cognition and Instruction, 10*(2–3), 249–260.

Chi, M. T. H., Slotta, J. D., & de Leeuw, N. (1994). From things to processes: A theory of conceptual change for learning science concepts. *Learning and Instruction, 4*(1), 27–43.

Church, R. B., & Goldin-Meadow, S. (1986). The mismatch between gesture and speech as an index of transitional knowledge. *Cognition, 23*(1), 43–71.

Cienki, A., & Müller, C. (2008). Metaphor, gesture, and thought. In R. W. Gibbs (Ed.), *The Cambridge handbook of metaphor and thought* (pp. 483–502). Cambridge: Cambridge University Press.

diSessa, A. A. (1993). Toward an epistemology of physics. *Cognition and Instruction, 10*(2–3), 105–225.

Dreyfus, B. W., Geller, B. D., Gouvea, J., Sawtelle, V., Turpen, C., & Redish, E. F. (2014). Ontological metaphors for negative energy in an interdisciplinary context. *Physical Review Special Topics – Physics Education Research, 10*(2), 020108 (11 pp.).

Dreyfus, B. W., Gouvea, J., Geller, B. D., Sawtelle, V., Turpen, C., & Redish, E. F. (2014). Chemical energy in an introductory physics course for life science students. *American Journal of Physics, 82*(5), 403–411.

Dreyfus, B. W., Sawtelle, V., Turpen, C., Gouvea, J., & Redish, E. F. (2014). Students' reasoning about "high energy bonds" and ATP: A vision of interdisciplinary education. *Physical Review Special Topics – Physics Education Research, 10*(1), 010115 (15 pp.).

Edwards, L. D. (2009). Gestures and conceptual integration in mathematical talk. *Educational Studies in Mathematics, 70*(2), 127–141.

Fauconnier, G., & Lakoff, G. (2013). On metaphor and blending. *Journal of Cognitive Semiotics, 5*(1–2), 393–399.

Fauconnier, G., & Turner, M. (2002). *The way we think: Conceptual blending and the mind's hidden complexities*. New York: Basic Books.

Fauconnier, G., & Turner, M. (2003). Conceptual blending, form and meaning. *Recherches En Communication, 19*(19), 57–86.

Forceville, C., & Urios-Aparisi, E. (2009). *Multimodal metaphor*. Berlin: Walter de Gruyter.

Goldin-Meadow, S. (1999). The role of gesture in communication and thinking. *Trends in Cognitive Sciences, 3*(11), 419–429.

Goldin-Meadow, S. (2003). *Hearing gesture: How our hands help us think*. Cambridge, MA: The Belknap Press of Harvard University Press.

Goldring, H., & Osborne, J. (1994). Students' difficulties with energy and related concepts. *Physics Education, 29*(1), 26–32.

Grady, J., Oakley, T., & Coulson, S. (1999). Blending and metaphor. In R. W. Gibbs & G. J. Steen (Eds.), *Metaphor in cognitive linguistics* (pp. 101–124). Philadelphia: John Benjamins.

Gupta, A., & Elby, A. (2011). Beyond epistemological deficits: Dynamic explanations of engineering students' difficulties with mathematical sense-making. *International Journal of Science Education*, *33*(18), 2463–2488.

Gupta, A., Elby, A., & Conlin, L. D. (2014). How substance-based ontologies for gravity can be productive: A case study. *Physical Review Special Topics—Physics Education Research*, *10*(1), 010113 (19 pp.).

Gupta, A., Hammer, D., & Redish, E. (2010). The case for dynamic models of learners' ontologies in physics. *Journal of the Learning Sciences*, *19*(3), 285–321.

Hammer, D. (1996). More than misconceptions: Multiple perspectives on student knowledge and reasoning, and an appropriate role for education research. *American Journal of Physics*, *64*(10), 1316–1325.

Hammer, D., Gupta, A., & Redish, E. F. (2011). On static and dynamic intuitive ontologies. *Journal of the Learning Sciences*, *20*(1), 163–168.

Hougaard, A. (2005). Conceptual disintegration and blending in interactional sequences: A discussion of new phenomena, processes vs. products, and methodology. *Journal of Pragmatics*, *37*(10), 1653–1685.

Jeppsson, F., Haglund, J., Amin, T. G., & Strömdahl, H. (2013). Exploring the use of conceptual metaphors in solving problems on entropy. *Journal of the Learning Sciences*, *22*(1), 70–120.

Jones, J. E. (1924). On the determination of molecular fields. II. From the equation of state of a gas. *Proceedings of the Royal Society A: Mathematical, Physical and Engineering Sciences*, *106*(738), 463–477.

Lakoff, G., & Johnson, M. (1980/2003). *Metaphors we live by.* London: University of Chicago Press.

Lakoff, G., & Johnson, M. (1999). *Philosophy in the flesh: The embodied mind and its challenge to Western thought.* New York: Basic Books.

Lancor, R. (2014). Using metaphor theory to examine conceptions of energy in biology, chemistry, and physics. *Science & Education*, *23*(6), 1245–1267.

Lawson, R. A., & McDermott, L. C. (1987). Student understanding of the work-energy and impulse-momentum theorems. *American Journal of Physics*, *55*(9), 811–817.

Machamer, P., Darden, L., & Craver, C. (2000). Thinking about mechanism. *Philosophy of Science*, *67*(1), 1–25.

McNeill, D. (1992). *Hand and mind: What gestures reveal about thought.* Chicago: University of Chicago Press.

Podolefsky, N., & Finkelstein, N. (2007). Analogical scaffolding and the learning of abstract ideas in physics: An example from electromagnetic waves. *Physical Review Special Topics—Physics Education Research*, *3*(1), 010109 (12 pp.).

Redish, E. F., Bauer, C., Carleton, K. L., Cooke, T. J., Cooper, M., Crouch, C. H., … Zia, R. K. P. (2014). NEXUS/Physics: An interdisciplinary repurposing of physics for biologists. *American Journal of Physics*, *82*(5), 368–377.

Reif, F. (1995). Millikan Lecture 1994: Understanding and teaching important scientific thought processes. *American Journal of Physics*, *63*(1), 17–32.

Roth, W. (2000). From gesture to scientific language. *Journal of Pragmatics*, *32*(11), 1683–1714.

Roth, W., & Lawless, D. V. (2002). Scientific investigations, metaphorical gestures, and the emergence of abstract scientific concepts. *Learning and Instruction*, *12*(3), 285–304.

Roth, W., & Welzel, M. (2001). From activity to gestures and scientific language. *Journal of Research in Science Teaching*, *38*(1), 103–136.

Scherr, R. E. (2008). Gesture analysis for physics education researchers. *Physical Review Special Topics—Physics Education Research*, *4*(1), 010101 (9 pp.).

Scherr, R. E., Close, H. G., & McKagan, S. B. (2012). Intuitive ontologies for energy in physics. *AIP Conference Proceedings*, *1413*(1), 343–346. doi:10.1063/1.3680065

Scherr, R. E., Close, H. G., McKagan, S. B., & Vokos, S. (2012). Representing energy. I. Representing a substance ontology for energy. *Physical Review Special Topics—Physics Education Research*, *8*(2), 020114 (11 pp.).

Slotta, J. D., & Chi, M. T. H. (2006). Helping students understand challenging topics in science through ontology training. *Cognition and Instruction*, *24*(2), 261–289.

Slotta, J. D., Chi, M. T. H., & Joram, E. (1995). Assessing students' misclassifications of physics concepts: An ontological basis for conceptual change. *Cognition and Instruction*, *13*(3), 373–400.

Smith, J. P., diSessa, A. A., & Roschelle, J. (1994). Misconceptions reconceived: A constructivist analysis of knowledge in transition. *Journal of the Learning Sciences*, *3*(2), 115–163.

Stivers, T., & Sidnell, J. (2005). Introduction: Multimodal interaction. *Semiotica*, *2005*(156), 1–20.

Watts, D. M. (1983). Some alternative views of energy. *Physics Education*, *18*(5), 213–217.

Wikipedia. (2014). *Energy level.* Retrieved May 22, 2014, from https://en.wikipedia.org/wiki/Energy_level

Wittmann, M. (2010). Using conceptual blending to describe emergent meaning in wave propagation. In K. Gomez, L. Lyons, & J. Radinsky (Eds.), *Proceedings of the 9th international conference of the learning sciences – Volume 1 (ICLS '10)*, Chicago (pp. 659–666). International Society of the Learning Sciences. Retrieved from http://dl.acm.org/citation.cfm?id=1854444&CFID=493068450&CFTOKEN=80132093

# Enacting Conceptual Metaphor through Blending: Learning activities embodying the substance metaphor for energy

Hunter G. Close[a] and Rachel E. Scherr[b]

[a]*Department of Physics, Texas State University, San Marcos, TX, USA;* [b]*Department of Physics, Seattle Pacific University, Seattle, WA, USA*

We demonstrate that a particular blended learning space is especially productive in developing understanding of energy transfers and transformations. In this blended space, naturally occurring learner interactions like body movement, gesture, and metaphorical speech are blended with a conceptual metaphor of energy as a substance in a class of activities called Energy Theater. We illustrate several mechanisms by which the blended aspect of the learning environment promotes productive intellectual engagement with key conceptual issues in the learning of energy, including distinguishing among energy processes, disambiguating matter and energy, identifying energy transfer, and representing energy as a conserved quantity. Conceptual advancement appears to be promoted especially by the symbolic material and social structure of the Energy Theater environment, in which energy is represented by participants and objects are represented by areas demarcated by loops of rope, and by Energy Theater's embodied action, including body locomotion, gesture, and coordination of speech with symbolic spaces in the Energy Theater arena. Our conclusions are (1) that specific conceptual metaphors can be leveraged to benefit science instruction via the blending of an abstract space of ideas with multiple modes of concrete human action, and (2) that participants' structured improvisation plays an important role in leveraging the blend for their intellectual development.

## Introduction

The general cognitive phenomenon of conceptual metaphor, recognized as significant in recent developments in cognitive science (Fauconnier & Turner, 2002; Lakoff & Johnson, 1999; Lakoff & Nuñez, 2000; Sfard, 1994), is also relevant to science education. We demonstrate that a conceptual metaphor of energy as a substance (Amin, 2009; diSessa, 1993; Duit, 1987; Falk, Hermann, & Bruno Schmid, 1983; Millar, 2005; Scherr, Close, Close, & Vokos, 2012; Scherr, Close, McKagan, & Vokos, 2012; Swackhamer, 2005) is particularly productive in developing understanding of energy transfers and transformations. We provide evidence that an embodied learning activity called Energy Theater engages learners with key conceptual issues in the learning of energy. In Energy Theater, each participant identifies as a unit of energy. Groups of learners work together to represent the energy transfers and transformations in a specific physical scenario. Objects in the scenario correspond to regions on the floor. As energy moves and changes form in the scenario, participants move to different locations on the floor and change their represented form. In previous work (Scherr et al., 2013), we have observed that Energy Theater supports the participants in engaging with key conceptual issues of energy, namely, disambiguating matter and energy, and theorizing mechanisms for energy processes; and that this engagement is supported by the material structure of the Energy Theater environment and the embodied action that it promotes. This environment was designed to be a metaphorical learning space with both a set of strict rules of engagement and with plenty of room for improvisation and emergent events and meaning. Energy Theater's metaphorical aspect is reminiscent of children's (or adult actors') imaginative play, in which a group agrees to pretend that things, people, and settings are not what they are in a literal sense. Energy Theater's rule structure is like that of a board game or team sport, in which rules are laid out, but the outcome is not determined by the application of these rules; instead it is achieved through an interplay between an adherence to the rules and many free, unplanned choices on the part of the participants. As we reflected on this combination of the three elements of metaphor, rules of engagement, and improvisation inherent to Energy Theater, it became apparent to us that a particular theory of cognition, *conceptual blending*, was very appropriate as a theoretical apparatus for understanding the workings of Energy Theater, especially in terms of the intellectual development of the participants. Thus, our present research question is: 'How can the perspective of conceptual blending account for the success of Energy Theater in connecting participants with key conceptual issues of energy?' To answer this question, we analyze new episodes of participant interactions, we add further evidence of participants' engagement with key conceptual issues, and we explore the use of concepts from *conceptual blending*, which we explain in further detail below in the section 'Vital Relations among Elements in Energy Theater'.

We justify the application of conceptual blending theory to our data in three ways: First, though Energy Theater was first designed without an explicit awareness of conceptual blending, its structure has always been formulated as a correspondence of elements in the concrete learning environment (e.g. people, ropes, hand signs) to elements in a separate 'physical scenario' space, with a spirit of the fusion of identity (e.g. 'I *am* a unit of energy', 'this bounded area on the floor *is* the pulley'), and with rules for human action (e.g. exactly one hand sign at a time) that are meant to communicate rules of the dynamics of energy (a unit of energy has exactly one form at a time). Second, the single overarching goal of all conceptual blending, according to Faucconier and Turner (2002, p. 322) is to 'achieve human scale'. We believe the achievement of human scale is a fundamental characteristic of Energy Theater. Thus the study of Energy Theater is in a reliable manner a study of conceptual blending itself. Third, by bringing the theory of conceptual blending to the analysis of our video records of Energy Theater, we illustrate its natural fit for understanding the meaning of events for participants.

As we analyze video episodes in detail, we take the theoretical perspective that the universal properties of an event or phenomenon emerge from the specifics of a particular case, rather than from the patterns that emerge across cases (Erickson, 1986). Our methodology is to identify video episodes in which learners engage with energy concepts in general and conduct detailed analysis to characterize the specific concepts with which they engage (Jordan & Henderson, 1995). A participationist theory of learning, in which learning is indicated by changes in speech and behavior, supports ethnographic analysis of learners' embodied interactions with each other (Lave, 1991; Sfard, 1998; Vygotsky, 1986; Wertsch, 2007) and the material setting (Hutchins, 1995; Jordan & Henderson, 1995; Nemirovsky, Rasmussen, Sweeney, & Wawro, 2011; Stevens, 2000). We conduct detailed analysis using conceptual blending theory to build plausible causal links between specific features of Energy Theater and the conceptual engagement that we observe (Maxwell, 2004a, 2004b; Salmon, 1998). The novel contribution of this work is to account for energy learning in terms of conceptual blending theory.

## Design of an Embodied Learning Activity Based on the Substance Metaphor for Energy[1]

### Embodied Cognition Perspective

In an embodied cognition perspective, all abstractions are understood in terms of basic sensory-motor experiences such as object permanence and movement (Lakoff & Johnson, 1999; Lakoff & Nuñez, 2000). Ideas such as time are expressed with embodied metaphors: for example, we might say that we are 'halfway through' the year, as though a year had spatial extent and we were moving relative to it. Human use of embodied metaphors is natural, unconscious, and pervades our talk; we often express conceptualizations of events, activities, emotions, ideas, and so on as being entities or substances. Embodied metaphors are often especially evident in the verbs and prepositional phrases used together with the terms of interest. For example, to say someone is 'in trouble' or 'close to graduation' conceptualizes these states as being locations, and to say that someone 'got an idea' or 'has a headache' poses these attributes as being possessions (Lakoff & Johnson, 1999). Grammatical indicators such as these can identify learner ontologies for energy—the kinds of things that people think of energy as being. Gestures and other bodily actions can also indicate ontologies (Close & Scherr, 2012; Scherr et al., 2013). Influential research in cognitive science has demonstrated that ontological categorization is key to understanding physics concepts (Chi, 2005; Chi & Slotta, 1993; Slotta & Chi, 2006).

## The Substance Metaphor for Energy

Certain statements pose energy as being a substance-like quantity—a kind of 'stuff'—and objects as being containers that can have such stuff in them:

> The gas *has* energy.

> Where did the energy *in* the gas *come from?*

Statements that implicitly treat energy as a substance are ubiquitous in physics textbooks and the words of famous physicists. Even statements that carefully avoid any explicit characterization of energy as anything other than an abstract numerical quantity use an implicit substance metaphor:

> We now introduce a third type of energy that a system can *possess* (Serway & Jewett, 2007).

> Thus, the flying duck *has* a kinetic energy of 6.0 J (Halliday, Resnick, & Walker, 2008).

> … When we *put* energy *into* the gas its molecules move faster and so the gas gets heavier (Feynman, Leighton, & Sands, 1969).[2]

This imagined substance is not a material fluid; rather it is a 'quasi-material' substance, one that includes certain properties of material substances (e.g. localization and conservation) but not others (e.g. mass, volume, viscosity). The substance metaphor for energy has limitations (Amin, 2009; Duit, 1987): it suggests that energy is not purely a mathematical quantity (Arons, 1965; Feynman et al., 1969; Warren, 1982, 1986), it does not support a concept of negative energy (Dreyfus et al., 2014), it does not include energy degradation or dissipation (Daane, McKagan, Vokos, & Scherr, 2015; Daane, Vokos, & Scherr, 2014), and it does not admit energy's frame-dependence or its delocalization in quantum mechanics, among other limitations (Duit, 1987). We have selected a substance metaphor for energy as a primary focus of our instruction because of its advantages for teaching conservation, transfer, and flow (Brewe, 2011; diSessa, 1993; Duit, 1987; Falk et al., 1983; Millar, 2005; Scherr, Close, McKagan, et al., 2012; Swackhamer, 2005). A substance metaphor supports the following features:

> *Energy is conserved.* This key feature is a primary advantage of the metaphor.

> *Energy is localized,* i.e. it is associated with a spatial location, even if spread out.

> *Energy is located in objects,* which are metaphorically represented as containers for energy.

> *Energy can change form.* As a material substance can change form (e.g. when a ball of clay is remolded, or when water freezes or boils), energy also can be understood to change in appearance or presentation while remaining fundamentally the same. Forms, in our

model, are categories of evidence that energy is present or changing, and thus an important means of connecting a unified energy concept to a variety of observable phenomena (McKagan, Scherr, Close, & Close, 2012).

*Energy is transferred among objects* and *energy can accumulate in objects.* Flow corresponding to a conserved quantity (i.e. a quantity subject to the continuity equation) is a key concept in physics, appearing here as energy transfer (and elsewhere as mass transfer, charge transfer, and momentum transfer, in both classical and quantum mechanical contexts).

These features constitute a powerful conceptual model of energy that may be used to explain and predict energy phenomena.

*Energy Theater*

Energy Theater is a learning activity designed to embody the substance metaphor for energy (Scherr, Close, Close, et al., 2012). In Energy Theater, each participant identifies as a unit of energy that has one and only one form at any given time. Groups of learners work together to represent the energy transfers and transformations in a specific physical scenario (e.g. a refrigerator cooling food or a light bulb burning steadily). Participants choose which forms of energy and which objects in the scenario will be represented. Objects in the scenario correspond to regions on the floor, indicated by loops of rope. As energy moves and changes form in the scenario, participants move to different locations on the floor and change their represented form. The rules of Energy Theater, which are presented explicitly to participants, are:

Each person is a unit of energy in the scenario.

Regions on the floor correspond to objects in the scenario.

Each person has one form of energy at a time.

Each person indicates his or her form of energy in some way, often with a hand sign or iconic movement.

People move from one region to another as energy is transferred, and change hand sign as energy changes form.

The number of people in a region or making a particular hand sign corresponds to the quantity of energy in a certain object or of a particular form, respectively.

Examples of forbidden actions would be:

A person identifies as an object rather than a unit of energy.

A person identifies as the energy of a particular object (rather than as a unit of energy that happens to be in the object at a particular moment in time).

A region on the floor is designated a form of energy (such as kinetic energy), rather than an object.

A person is energy for only part of the activity and then sits down or otherwise leaves the scenario.

A person remains energy but moves to an unmarked region on the floor, corresponding to nowhere in particular.

The verbal narrative of the physical processes in the scenario is inconsistent with the embodied 'narrative' in the Energy Theater space, e.g. a box is said to move at constant speed but the number of units of kinetic energy in the region corresponding to the box does not remain constant.

Examples of extemporaneous moves that are beyond the scope of the rules and can contribute to emergent meaning would be:

A person moves when others think that person should stay, causing an interruption in action and a discussion of a conceptual issue.

A hand sign corresponding to a form of energy causes confusion due to its visual similarity to another sign.

By chance, a group has an odd number of people and therefore cannot divide the total energy into two equal parts, causing people to attempt to account for an energy unit that is 'left over.'

People imagine and recount amusing consequences for the physical system if a mistaken enactment of Energy Theater were in fact correct, or if energy that is neglected in the enactment, but which is understood to be present in the scenario, were in fact absent. For example, people might joke that an object with no represented thermal energy is at a temperature of absolute zero, or a man pushing a box dies when his energy is exhausted in the enactment.

In designing Energy Theater, we have sought to specifically harness the affordances of the energy-as-a-substance metaphor by developing a representation that embodies that metaphor. Since one of the most basic experiences of substances is that of object permanence, we developed a representation in which energy is explicitly shown as being an object or objects; and since a particularly cognitively compelling sense of permanence might be attached to the self, and use of the human body might have special significance for learning, we developed a representation in which people identify as units of energy. Energy Theater is thus embodied in two separate senses: it makes explicit use of a particular experientially grounded metaphor (energy as a quasi-material substance), and it uses the human body to symbolize physical entities (Stevens, 2012). A variety of other embodied learning activities have been developed in which the body represents mathematical entities (Touval & Westreich, 2003), molecules (Ross, Tronson, & Ritchie, 2008), electrical charges (Manogue et al., 2001; Singh, 2010), celestial bodies (Morrow, 2000; Reinfeld & Hartman, 2008; Richards, 2010), computer science entities (Begel, Garcia, & Wolfman, 2004), components of a dynamic system (Colella, 2000; Resnick & Wilensky, 1998), cellular processes (Chinnicci, Yue, & Torres, 2004; Wyn & Stegnik, 2000), and even literary devices (Zimmerman, 2002).

## Conceptual Blending Theoretical Perspective on Embodied Learning Activity

We understand Energy Theater as a blend of two spaces, in the manner described by the theory of cognitive or conceptual blending (Fauconnier & Turner, 2002; hereafter,

'F&T'). According to conceptual blending theory, blends are ubiquitous, sometimes spectacular but most often unnoticed, and very useful for human thinking and communication. (Energy Theater is probably not one of those everyday, usually unnoticed blends.) A blend always involves at least two input spaces and creates a blended space that incorporates some elements and relations from the input spaces. For example, 'Weird Al' Yankovic has made a successful career as a satirical musical artist by taking (in many, but not all, cases) just the music of a popular song (e.g. 'Happy', by Pharrell Williams), creating new lyrics with a center of meaning in another domain (self-serving, tasteless behavior) and combining them to make a new song ('Tacky'). In the case of Energy Theater, the two input spaces are (1) the literal learning space with people, a floor, ropes for bounding regions on the floor, and other incidental environmental features (such as furniture), and (2) the space of the physical scenario (e.g. a box sliding down an incline while slowing down), which may be demonstrated physically or is sometimes only imagined. The physical scenario space may also be modeled as a blend of two spaces: the concrete, observable space of objects, and the abstract, imaginary space of energy. We do not pursue an analysis of this object-energy blend in this article, though we recognize that for most participants this blended 'physical scenario' space is surely hazy and incomplete. One major purpose of the Energy Theater blend is to clarify the internal relational structure of the physical scenario space; that is, a major purpose is to teach physics.

*Vital Relations among Elements in Energy Theater*

Vital relations in conceptual blending are those fundamental relationships between any elements in any of the spaces (F&T, p. 93). Vital relations between elements A and B could be of various forms: A causes B, A represents B, A is a part of B, A happens before B, A is above and slightly to the left of B, A looks like B, A is B, etc. If the relationship is between elements within one space, it is called an inner-space vital relation; if it is between elements in two different spaces, it is called an outer-space vital relation. For instance, we understand a photograph of a person to be a representation of that person; the vital relation is representation. Representation can compress in a blend to an inner-space vital relation of uniqueness, in which we treat the photograph and the person as one and the same, perhaps saying, 'Look at the expression on your face!' An exhaustive set of vital relations and some descriptions of their dynamics are given in F&T, especially chapters 6 and 16.

Using the language of *vital relations*, we understand the 'people' input space to have the following structure that is imported to the blended Energy Theater space: People are *elements* in the space and are interchangeable members of the *category* 'participant'. The space also contains 'region' elements that are marked by closed loops of rope; these bear a special *spatial* inner-space vital relation of containment to the people: each person is either inside or outside any given rope. Each person makes a single hand sign at a time, and this hand sign is understood as a changeable *property* of the person. Actions in the people space also gain meaning by their *time* ordering—whether event A occurs before, after, or at the same time as event B is often important for people in general, and it is also important in Energy Theater. Despite being members of a category, and being able to function as interchangeable, each person also is easily distinguishable, if their distinguishability can provide some cognitive advantage. Indeed, each person has a flexible self-*identity* relation that can be engaged or disengaged; a person can shift identity from one interaction sequence to another to disconnect actions and prevent them from gaining meaning from each other by sharing a context, or a person can maintain an identity (especially when acting as a single energy unit) to express *change* through different sequential actions. This resource for human interaction is easily seen in the act of telling a story: we understand when the storyteller recalls a conversation between Harry and Sally that some utterances belong to Harry and some to Sally, and that some other utterances by the storyteller belong entirely outside the conversation between Harry and Sally. We present examples of the dynamics of identity in Energy Theater below in the section 'Energy Learning through Embodiment of the Substance Metaphor'.

*Inherited Structure from Input Spaces*

Part of the utility of blends is that the input spaces can provide structure to the blend. In the example of Yankovic's satiric songs, the blended song inherits its musical structure from the parodied song. In Energy Theater, the 'people' space brings conceptual structure to the blend partly through its material structure (Hutchins, 1995, 2005), including the facts that people are conserved, each person has a location, and, when counting people, the whole is automatically the sum of the parts. When the people become units of energy, these structures are inherited from the 'people' input space into the blended space, so that units of energy are conserved, located, and easily summed.

The 'people' input space provides structure to the activity also through the structure of existing social resources (Goodwin, 2000; Greeno, 1998; Hutchins, 1995). Part of the structure is provided by the explicit rules of Energy Theater, which perhaps seem arbitrary at first. However, this arbitrariness can still be experienced as culturally coherent in that many games and puzzles are presented initially as a set of arbitrary rules whose value or meaning is discovered through their application. Other aspects of social structure for Energy Theater provided by the people input space come from participants' experience managing disagreement within groups of people. The social structure includes both the explicit rules of Energy Theater and other tacit rules of a broader culture, like the tendency to work out disagreements through debate and compromise. Thus, in the blended space, units of energy move around and transform and also debate with each other about how to move around and transform.

The scenario input space brings less structure, or a less reliable structure, to the blend, since it is sensitive to the specific physical scenario and to the media by which the scenario is apprehended (e.g. experiment, common memory, quasi-theoretical simulation, or intuitive prediction), and to what the participants know or believe about the scenario. For example, participants may think the cart speeds up, or the ice water maintains a constant temperature as it melts, or may have other partial information and understanding that can help to guide and constrain the participants' solution to the puzzle of how to depict the energy. Hence many details of the emergent structure of the Energy Theater blend are unpredictable. The strict rules for symbolic engagement in the blended space are intended in the design of the activity to result through their repeated application in a clear, shared understanding of various scenario input spaces.

Regardless of the changing conditions of the physical scenario input space, some particular 'outer-space' vital relations are likely to hold: First, almost inevitably the relation of perceptible *similarity* between participants in the people space and the objects in the scenario space will lead participants to form an erroneous *uniqueness* relation between themselves and some object (i.e. someone becomes an object instead of energy). Conversely, initially the energy in the scenario space is likely 'perceived' (or not perceived at all) as *dissimilar* to participants in the people space. *Analogy* and *disanalogy* are outer-space vital relations that F&T claim usually compress into *similarity* and *dissimilarity* in a blend; that is, those things that are understood to be abstractly alike (or not alike) are reconceived to appear alike (or not alike). Therefore, it seems reasonable that in Energy Theater, participants must effortfully remember that what appears alike in the blend does so despite the fact that it is not meant to be analogous in the relation between elements in the input spaces. Second, in a move that is neither prescribed nor forbidden by the rules of Energy Theater, participants regularly compress *part-whole* relations into *uniqueness*; participants often show through their own actions what they believe many, or an indeterminate number of, energy units would do. The part-whole relation is shown to be compressed in the blend through the typical *lack* of discussion about how exactly to scale up the action from one person to seven people to a thousand energy units.

Another crucial component of the conceptual structure of the blended space is the fact that an Energy Theater enactment is by nature a group product. Each person in the group has direct authority over the behavior of one unit of energy, the individual self; but the product of the group's work is one coherent Energy Theater enactment, regardless of the number of opinions in the group about what should happen to the many energy units. As when a barbershop quartet sings a chord, one participant who makes a distinctive contribution changes the whole result for everyone. To whatever degree each person is invested in a particular proposed global solution to an Energy Theater problem, that person is invested to the same degree in persuading others to see the value in the proposed solution, since any global proposal requires others' willful cooperation if it is to be part of the final performance or solution. The result is a high intensity of negotiation of meaning (Scherr et al., 2013).

## Summary

In summary, Energy Theater cognitively blends learners and energy together to create an embodied problem-solving and concept-exploring space. The purpose of the specific blend is to create a situation that stimulates intense negotiation of meaning about energy. The negotiation of meaning arises mostly naturally, in the sense that the interaction of human need for meaningful experience with the structure of the activity is sufficient to call for the negotiation; little direct instructor intervention is required after participants have understood the basic structure for activity. The fact that this negotiation of meaning is officially approved as on-task behavior promotes genuine participation and gradual transformation of the learners. The specific character of the negotiation is authentic to the broader community of practicing physicists in the sense of using disciplined imagination of the dynamics of hypothetical entities (Ochs, Gonzales, & Jacoby, 1996).

## Research Methodology

*Methodological Perspective*

The use of rich records of naturally occurring activities as evidence of learner knowing promotes and supports a socio-cultural view, in which learning is a process that shows in what participants do and say together (Sfard, 1998, 2007). For this view of learning, ethnographic perspectives are naturally relevant (Erickson, 2004; Mcdermott, Gospodinoff, & Aron, 1978; Schegloff, 1997). We identify with the interpretive tradition (Erickson, 1986), in which the phenomena of interest for learning are the meaning of activities for the participants. This perspective asserts that participants create meaningful interpretations of physical and behavioral occurrences; that they take action based on their interpretations, that is, interpretations are causal; and that these interpretations are often invisible to participants, who treat their interpretations as reality (Denzin & Lincoln, 2005). A primary function of the ethnographic researcher in this tradition is to 'make the invisible visible' (Goodwin, 1994): to describe the implicit social and cultural organization that shapes the participants' activity (Anderson-Levitt, 2006).

*Participants*

Our data consists of videotaped episodes of teachers in a professional development course analyzing the energy dynamics of specific real-life physical scenarios. The episodes are from video records of professional development courses for K-12 teachers offered through Seattle Pacific University as part of the Energy Project, a six-year, NSF-funded project to develop and study teacher practices of formative assessment in the context of energy teaching and learning. Teachers participate primarily in order to gain content understanding of energy for themselves, and secondarily to translate that content understanding into classroom activities that are aligned with national science learning standards (National Research Council, 2012; NGSS Lead States, 2013) and other constraints. Participating teachers teach in a wide range of situations, with an enormous variety of populations, material resources, institutional constraints, and expectations for science learning. For this reason, the professional development experience does not provide teachers with a curriculum or script to enact in their own classrooms. Instead, project team members work to encourage and support each teacher in creatively applying sound conceptual understanding to their own circumstances. Some teachers elect to use Energy Theater in their own class-rooms (Daane, Wells, & Scherr, 2014).

## Data Collection

Energy-centered professional development courses offered by the Energy Project are documented with video, field notes, and artifact collection (including photographs of whiteboards, written assessments, and teacher reflections). In each course, teachers are grouped into 4–8 small groups, and two groups are recorded daily. As researcher-videographers document a particular course, they take real-time field notes in a cloud-based collaborative document, flagging moments of particular inter-est and noting questions that arise for them in the moment. Later, the researcher-videographers or other members of the Energy Project identify video episodes to share with a research team. We use the term 'episode' to refer to a video-recorded stretch of interaction that coheres in some manner that is meaningful to the partici-pants (Jordan & Henderson, 1995). These episodes are the basis for collaborative analysis, development of research themes, literature searches, and the generation of small or large research projects.

## Episode Selection

The episodes in this paper were selected from an Energy Theater enactment initially observed by author R.E.S. in the summer of 2013. In this enactment, participants negotiate and perform Energy Theater for adiabatic compression of a gas. R.E.S. highlighted this particular Energy Theater enactment on the basis of audio-visual clarity, sustained learner engagement with a physical scenario, and appropriate implementation of Energy Theater, that is, the participants mostly followed the rules specified in section 'Energy Theater'. The enactment analyzed in this paper is not the only enactment with these features, and we do not present evidence that it is a representative enactment—that is, we do not present evidence that most other enactments have the same features (though our experience suggests that many of them do). Rather, we put forward this enactment as a *case* of Energy Theater: an instantiation through which we may identify universal features of Energy Theater that are evident in the concrete details of its practice. We expected that this enactment would help us to identify some of the ways in which conceptual blending is manifested in the activity.

## Analysis

After identifying this enactment as one likely to contribute to answering our research question, each author watched the video multiple times, creating a detailed narrative of events as well as a sketch transcript focusing primarily on speech. On the basis of the narrative, sketch transcript, and multiple viewings, two episodes were selected in which learners engage with energy concepts and with one another as they construct an understanding of energy. These episodes were isolated and transcribed in greater detail, including transcription of embodied actions. Claims were developed that responded to the research question: 'How can the perspective of conceptual blending account for the success of Energy Theater in connecting participants with key conceptual issues of energy?' We respond to this question with interpretive video analysis combined with theoretical study of conceptual blending as evidenced in human interaction. We conduct detailed analysis to describe specific features of Energy Theater and the learning events that we observe in the language of conceptual blending. In line with our participationist perspective, learning events include those in which learners' talk moves toward expert use of disciplinary language.

## Energy Learning through Embodiment of the Substance Metaphor

In the episodes below, we show how Energy Theater supports a group of teachers in learning about energy. Specifically, we show how teachers' embodied engagement in the blended Energy Theater space supports them in grappling with key conceptual issues in energy. The physical scenario being considered in the episodes below involves the adiabatic compression of an ideal gas. The scenario was communicated to the teachers by the instructor of the professional development course (author R.E.S.) primarily through a projected display of a PhET simulation (Weiman, Adams, & Perkins, 2008) on the classroom screen rather than through a verbal description. In the simulation 'Gas Properties', a man (whom the teachers sometimes called 'the man' or 'Scuba Steve') pushes with his whole body on a movable wall of a container of gas, decreasing the volume of the container (see Figure 1).

The gas is depicted only in terms of its constituent particles (which the teachers sometimes called 'purple balls' because of their appearance in the simulation). In steady state, the balls move around inside the container, colliding very often with each other and the container wall. As the man moves the wall, the balls are seen to move faster. A simulated red alcohol thermometer is attached to the container and shows a rise in the level of the alcohol and an increase in the numerical temperature reading. Teachers were simply challenged to produce an Energy Theater 'solution' to the scenario, which is a routine to which they had become accustomed, having had a full two-week intensive course ('Energy I') the previous summer (2012) and one full week of instruction using Energy Theater so far in Summer 2013 in a second-year course ('Energy II'). The scenario was their seventh Energy Theater challenge so far that week. In the episodes below, seven teachers (pseudonyms Dan, Andy, Scott, Elaine, June, Denise, and Sally) begin to negotiate such a solution. Before the episodes described below begin, these teachers had already laid out three rope loops in a linear sequence of adjacent loops, each in a generic rounded rectangular shape. In the analysis below, we refer to ropes 1, 2, and 3, while the teachers refer them in blended fashion with the labels 'man', 'wall', and 'balls'.

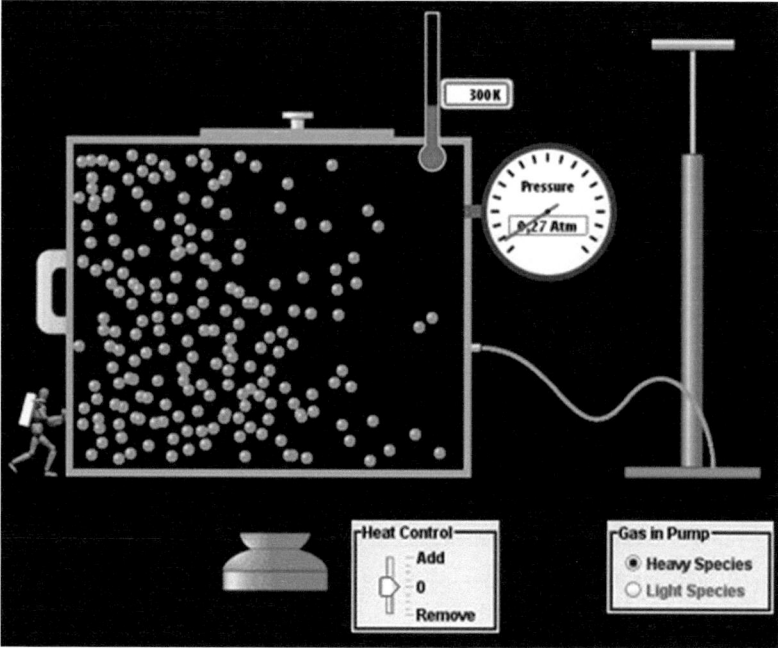

Figure 1. Screenshot from 'Gas Properties', an interactive simulation by PhET Interactive Simulations, University of Colorado, Boulder. http://phet.colorado.edu.

When 'doing Energy Theater' as an instructional activity, groups of participants tend to spend most of their time arguing over different proposals for enactments of Energy Theater rather than perfecting their execution of a sequence of coordinated actions. Indeed, teachers seem to take up Energy Theater primarily as a puzzle-solving tool in an intense group negotiation that hybridizes spoken discussion and argumentation with symbolic embodied action. During the planning, the effect of using the rule structure of Energy Theater is to constrain and regiment the group's thinking and argumentation with a particular theoretical perspective on energy itself.

*Distinguishing among Energy Processes*

In the following episode, teachers distinguish distinct energy processes (transfers and transformations) that are potentially important to the energy dynamics of the scenario (see Figure 2).

Participants make this progress by engaging with the symbolic material structure of Energy Theater in a variety of basic manners: simulating energy transfer with body locomotion, using the space as a reference frame for gesture, coordinating movement with speech, and enforcing rules of interpretation. In all of these ways, the inherited structure of the blend shapes and promotes learners' understanding.

Dan initiates the group discussion by proposing a particular Energy Theater solution in words and by 'walking it through'. After his proposal, the group considers

Figure 2. Participants negotiate Energy Theater for adiabatic compression of a gas. In this episode, participants distinguish distinct energy processes that are potentially important to the energy dynamics of the scenario

the possibility of an energy transformation that was not an explicit part of Dan's proposal. The episode is about 1.5 min long (see Supplemental data).

At the beginning of this episode, most participants are standing around the perimeter of the set of ropes. Only two participants are standing inside the ropes: Dan is inside rope 1, and Andy is inside rope 3. The transcript that follows annotates participants' speech with a description of the embodied action that accompanies it. The '==' symbol indicates an unbroken turn at talk, overlapped by another speaker (Table 1).

The intellectual progress of this episode is for participants to distinguish among related, but distinct, energy processes that comprise the energy dynamics of adiabatic compression. In particular, the participants distinguish a transfer of energy from man to wall to gas ('one pathway is kinetic energy man, kinetic energy wall, kinetic energy balls') from a transformation of energy within the gas ('the moving purple ball inside the gas is transforming into heat'). This distinction is only partially articulated at this time; for example, June refers to the gas particles transforming (rather than the energy of the gas particles), and her use of the term 'heat' is not canonical (Kraus & Vokos, 2011; Scherr et al., 2013; Scherr & Robertson, 2014). Nonetheless, the distinction between energy transfer to the gas and energy transformation within the gas is crucial to the analysis of the scenario.

When Dan initially proposes the rudiments of his Energy Theater solution, he steps through the regions and coordinates his talk (e.g. 'kinetic energy, wall') with his arrival in the corresponding region and points downward at each region, enacting the Energy Theater blend by establishing the meaning of a representative unit of energy passing

Table 1.   The transcript in Table 1 annotates participants' speech with a description of the embodied action that accompanies it

| Speaker | Speech | Embodied action |
|---|---|---|
| Dan | So one pathway is kinetic energy man kinetic energy wall kinetic energy [1 sec] balls | Stands in rope 1, points down to rope 1. Walks through rope 2, points down to rope 2. Arrives at rope 3 and points down to rope 3 |
| Andy | Mmm-hmm | |
| Scott | Yeah. I think there's one other kind a tricky thing is because the size of the box changes, we are compressing the air so the air is now running into itself much more often. So it's a little more complicated than just adding some kinetic energy—we're not thumping == | 'Squeeze' gesture: Positions hands as though he is holding a package of air and moves them together Dan walks backwards to return to loop 1. Finger-flicking gesture directed at screen |
| Elaine | Well == | |
| Scott | == a few of the molecules | |
| Elaine | == kinetic turns to thermal | |
| Scott | Well yeah, definitely! | |
| Andy | | Steps backwards and exits rope 3 |
| June | Can this be a separate tr—a separate transformation ... so | Approaches rope 3 |
| | if she—the—not focusing on that pathway but the purple ball, the moving purple ball inside the gas is transforming into heat | Turns back to her right and points to Dan in rope 1. Hand thrusts forward, toward rope 3 |
| | I guess it would have to be a kinetic-kinetic then thermal | Steps back |
| | Like, can that transformation happen only here Can that pathway exist only inside. The. Molecules. [5 sec] 'Cause we've always had like a starting | Steps forward into rope 3 Points down to rope 3 Pause, steps back and out Holds left palm out to Dan's position in rope 1, whistles, flips left palm around and sweeps to her left, toward rope 3 |
| | I feel like something's happening THAT pathway | Points to rope 3 Sweeps pointer from rope 1 to rope 3 |
| | There's also something happening with the moving balls inside here. | Steps back into rope 3 |
| Dan | There's something internal | Steps forward into rope 2 |
| June | [inaudible] | |
| Scott | We have some heat energy that we're starting off with == | |
| Dan | So | Steps into rope 3 |
| Scott | == they're definitely moving, we're above Absolute zero | |

*(Continued)*

Table 1.   Continued

| Speaker | Speech | Embodied action |
| --- | --- | --- |
| Dan | But what if we were moving about and we == | Performs repeated double high-fives with Andy |
| June | YES | |
| Dan | == did this for pressure | |
| Andy | But | Closes fingers of high-five hands |
| Denise | Nooo | |

from one object to the next in sequence. It might seem that his string of speech has a definite meaning on its own and that his body movement is superfluous; he describes a 'pathway' and states the form of energy and the location in each phrase. However, in a different context his words might have been interpreted as describing the presence of kinetic energy in each object (man, wall, and balls) simultaneously, rather than sequentially as a transfer. Thus, his deliberate walkthrough helps to establish the structure of the blend.

Dan's actions and words communicate vital relations that are internal to the blended space: *space, time, identity*, and *property* (form of energy). In fact, they seem even to communicate nothing besides these vital relations. The grammatical and pro-sodic structure of Dan's speech is shaped in part by the metaphorical space of Energy Theater and the 'questions' it 'asks' its users (Scherr, Close, McKagan, et al., 2012): Where will you stand? Where will others be at the same time? Where will you go next? What form will you be? and so on. As the space asks these questions, Dan's speech is structured to provide those answers. This idea that learner activity is substantially shaped by the forms of representation has been both argued generally and shown empirically in the case of learning of physics when algebraic structures for thinking and communicating are replaced with computer code (Sherin, 2001). Dan's concise solution description suggests that he thinks that his solution is correct in its basic form; he ends the sequence on the word 'balls' with a drop in pitch that signals the end, after which he pauses briefly and solicits agreement. Dan's solution is presented as though no other significant components will be necessary as it is adapted to a full-fledged solution involving several actors (though perhaps the timing might need to be fine-tuned). Thus, a single energy unit (Dan) becomes all the energy in the scenario for a short period, such that Dan is all the energy. Using the parlance of blending: The part-whole vital relation between the single energy unit and all the energy in the scenario temporarily compresses in the blend to unique-ness (F&T, p. 113); as a result of the compression, the single unit of energy *is* all the energy. June's concern is primarily with understanding if and how there is an energy transformation happening inside rope 3, or 'inside the molecules', perhaps to help explain how kinetic energy becomes thermal energy. June, like Dan, points to elements of the Energy Theater space and refers to them as though they were the objects they represent (man, molecules, etc.), which indicates that the intended blend between the 'people' space and the 'scenario' space has been successful. Additionally, June's speech contains several missing pieces that are filled in through her embodied reference to the blended Energy Theater space, which incidentally placed her literally at the center of the group's interaction. Therefore it is plausible that June's meaningful contributions to the discussion are enabled through her engagement in the blend, and that without support from the blend (as it appears in the material structure of the environment), her point might have been overlooked by the group. June's engagement with the Energy Theater space in this episode is not as plainly ordered as Dan's: she steps in and out of the ropes; she interrupts her own speech to refer back to Dan's proposal and to rearticulate her concerns; she uses a mixture of gestures and locomotion to organize and express her thinking. However, June's actions assemble into a meaningful whole that is assisted by the spatial structure of the ropes, their metaphoric correspondence with objects, and by the vapor trail of Dan's proposal in her memory.

In contrast to the manners of participation of Dan and June, Scott speaks and gestures clearly and authoritatively, but in ways that are mostly not coordinated with the Energy Theater space. His body remains in one location outside and next to rope 3, and his metaphoric gestures (squeezing the gas, flicking the molecules) do not refer to the rope- objects, people-energy units, or to anyone else's motion. His ideas are potentially relevant to the group's discussion; the collisions between gas molecules can be understood as an energy transfer mechanism between molecules, and perhaps even as an energy transformation mechanism, as the means by which ordered kinetic energy becomes disordered and thereby reclassified as thermal energy. However, the response of the group to Scott's ideas is tenuous, perhaps because they find the meaning unclear as the ideas are not expressed in terms of the group's common metaphor.

*Disambiguating Matter and Energy*

The episode above not only shows teachers distinguishing among energy processes in the scenario, but also disambiguating matter and energy. When Dan attempts to incorporate the idea of collisions between molecules with a 'double-high-five' inter-action between himself and Andy, Andy and Denise object, probably because Dan lapses into identifying with the particles instead of the energy. In this case, this lapse is probably attributable to the visual salience of the 'purple balls' representing the gas particles in the displayed simulation. The participants and the balls look alike in certain ways: there are many of them, they are moving, they are inside a bounded region, etc. However, their similarity of appearance is accidental, and not the result of human cognitive activity deliberately constructing their similar appearance in order to communicate a deeper, more abstract likeness, as is the case in the relationship between persons and energy units in the design of Energy Theater. In other words, the participants and the balls have an outer-space vital relation of *similarity*, but the similarity is not a compression from *analogy*. Many Energy Theater participants have been observed to treat matter and energy interchangeably at first

(Scherr et al., 2013). Energy Theater contributes to the disambiguation of matter and energy by encoding a distinction between material objects and energy: energy (represented by participants) is located in objects (represented by areas demarcated by loops of rope). The energy in a scenario is clearly distinct from the objects (i.e. participants stand within, but are not mistaken for, areas inside loops of rope). The activity of developing an Energy Theater enactment for this scenario causes the group to attend to distinctions between matter and energy. As explained above in the section 'Research Methodology', we understand the process of participants enforcing the rules of Energy Theater as a major contribution to the learning of physics that is achieved during Energy Theater.

*Identifying Energy Transfer*

In this next episode, which follows soon after the episode above, one teacher (Sally) leads a discussion (see Figure 3) establishing that since the temperature of the gas increased during compression, there must have been a transfer of energy to the gas. Her model contrasts with June's model, which attributes increased temperature to transformation of internal energy in the gas. Sally accomplishes this by adding the vital relation of *identity* to units of energy. This episode is almost 4 min long (see Supplemental data) (Table 2).

The theme of this episode is captured by Sally's climactic question, 'Where is the energy coming from?' The question is clearly in accord with the Energy Theater

Figure 3. Participants negotiate Energy Theater for adiabatic compression of a gas. In this episode, participants establish that since temperature of the gas increased during compression, there must have been a transfer of energy to the gas

Table 2.  Participants' speech with a description of the embodied action that accompanies it

| Speaker | Speech | Embodied action |
|---------|--------|-----------------|
| Sally | So here's my question, 'cause this is the one that I'm like y'know, I'm NOT productively stupid with this one | |
| Group | Laughter | |
| Sally | This—the speed of the molecules is constant right now, it's 300 K, right? | Turns to the screen and points at it |
| Andy | Mmm-hmm. | |
| Sally | 'Cause that's what this is, a measure | |
| Scott | Yes. | |
| Elaine | Yeah, it's an average, but == | |
| Sally | Right, the temperature? | |
| Elaine | ==it's a constant average. It's an indicator | |
| Sally | Are we agreeing … When we squish the box the temperature goes up so the speed of those molecules has changed. Is that correct? | 'Squeeze' Raises hand vertically with palm face down Petting motion—palm of right hand mostly vertical, also perhaps like a gentle 'halt' gesture |
| Group | Yes. It's true. | |
| Sally | Stop. That's what I … But then once the box stops squeezing, they stay the same speed after that Correct? | 'Halt' gesture: Holds single palm vertically Facing palms to show box 'Stay' gesture: Places palm facing forward and down |
| Group | | Nodding |
| Sally | Now. [Dramatically] Where is the energy coming from? Is that box losing heat. Energy. | 'Window-wiping' gesture: Two palms forward, moving out and down, followed by 'halt' 'Rodent' gesture: Holds hands together as though a rodent holding food Adapts 'rodent' to point to simulation Slaps own right cheek several times quickly, as if to reprimand herself for using the word 'heat' improperly |
| Elaine | In the, in the, in the volume change? | |
| Sally | Are we saying 'here, I have this molecule with all this kinetic energy and it's the same amount of energy,' right? Now, somewhere in the proces we said that heat energy is going to be produced Is heat energy being produced? | Steps into rope 3 Performs group's iconic gesture for kinetic energy, 'Choo-choo' gesture: bent arms pump like the connecting rods of a steam locomotive 'Inclusion' gesture: Spreads hands outward 'Choo-choo' 'Choo-choo' |
| Andy | Yes. | |
| Sally | So, is the heat energy the same as the kinetic energy? | 'Choo-choo' |

*(Continued)*

Table 2. Continued

| Speaker | Speech | Embodied action |
|---------|--------|-----------------|
| Scott | The | |
| Sally | That's myyyyy Right? | Slowly throws hands up into the air |
| Scott | OK, I think what we gotta think about is that—what—I'm going to change what I said earlier. 'Cause I was saying, and you know, just the wall moving is going to provide some, and the banging into each other is going to provide some. I actually, I'm going to change that. I think it's, because that wall is difficult to push | Sally steps out of ropes

'Passage' gesture: Even sweep of both arms toward rope 3

'Push' gesture: Lines up arms with edge of rope 3 as though to push on the side of the container and compress the gas |
| | Because of the pressure that's inside there, I think that it really is providing All of that heat energy, that added heat energy must come 'cause this is the only input of energy is this wall moving. | Diminished 'Squeeze'

Diminished 'Push'

Points to rope 2 'Receive' gesture: holds upward facing palms toward rope 2 'Passage' |
| | So that kinetic energy going in there and squeezing that—that — volume of air, that space, it's the only input of energy in, so it must be | 'Snowball' gesture: Two hands pack material into a small ball |
| Sally | It's giving it more kinetic energy to the molecules | |

metaphor: specifically, not only does energy come from somewhere, but it is of great importance to know where it came from. To prioritize this question is to prioritize the vital relations *space* (energy has location), *time* (energy events are sequenced), and *identity* (energy can be identified and tracked) in a particular pattern that expresses local conservation. The *identity* vital relation is implicated with Sally's use of the word 'the' in 'the energy'; her perspective assumes that energy can have an identity, and that it is meaningful to ask about 'this energy' versus 'that energy'. By adding the vital relation of *identity* to units of energy, participants are encouraged to search locally for mechanisms of transfer.

The follow-up question 'Is the heat energy the same as the kinetic energy' is of comparable importance within the metaphor, since the forms of energy should be understood clearly and distinctly in order to correspond with observational evidence. In particular, observing that a system is hot is not exactly the same as observing that it is fast. Thus, for the group to agree that heat energy (of the gas) and kinetic energy (of the molecules) bear a kind of identity relation requires them to recognize the transformation of the observation depending on its scale; a 'microscopic' observation would see the fast molecules, and a 'macroscopic' one would see a high temperature reading. In a manner similar to a conductor of an orchestra, Sally dramatically leads the group to consider her questions by eliciting the group's assent to various statements of phenomenological fact about speeds of molecules and temperature readings. Scott's response to Sally's questions is a mixture of phenomenological statements ('that wall is difficult to push') and gestures (squeezing and pushing) and ambiguous Energy Theater metaphor-style statements ('kinetic energy going in there and squeezing that volume of air'). Despite its lack of focus, however, Sally seems to use his input to conclude a clear statement of transfer of kinetic energy from the wall to molecules.

*Representing Energy as Conserved*

Interactions immediately following the previous episode establish energy as a conserved quantity, one that increases through a mechanism of flow and under a constraint of constant total amount, rather than through other means such as additional activity. This progress supports the group in theorizing about mechanisms of transfer and modeling energy as a discrete quantity. The group accomplishes this intellectual progress by negotiating different members' improvised variations on the Energy Theater blend (Table 3).

At the end of the previous sequence of action, Scott had suggested that energy is transferred from the wall to the molecules ('this is the only input of energy is this wall moving'). In the present sequence, Sally explores how this input is properly represented in the Energy Theater space. She tries out two different, and almost simultaneous, ways of imaging this energy input. First, she tries modifying the group's iconic gesture for kinetic energy (the 'choo-choo' gesture, simulating the pumping action of locomotive wheels with bent arms pumping forward and backward close to the body) to have faster arm movement. This proposal suggests a correspondence of speed (of molecules) with speed (of her arms). Sally's proposal is enacted, not declaimed; that is, it is not accompanied by any directly corresponding speech like 'Should I move my arms faster to show more kinetic energy?' Very shortly after this first proposal, she asks if her size should increase; she asks both with her words and with a gesture that encircles her body as though to trace out a larger person. This suggestion (increase in size) is a correspondence between speed and continuous spatial extent, as though the energy were more of a continuous mass than a multiplex of many individuals (Lakoff, 1987). Andy responds to Sally's question about size by with a definitive 'no' and prepares to act out the transfer of additional energy units to the molecules. Sally catches on ('So there's more of us in here?') before Andy or anyone else actually enters rope 3, representing the molecules. This conclusion is further affirmed verbally by Sally as she, Andy, and Scott act it out.

Table 3.  Participants' speech with a description of the embodied action that accompanies it

| Speaker | Speech | Embodied action |
|---|---|---|
| Scott | Yeah, and, and— | |
| Sally | But also at the same time, we've got this temperature going up, which I'm hearing people saying is a measure of an increase in heat energy. OK? So my question is | |
| | [chuckles] … 'Cause I can't see us | Steps into rope 3, steps backward into rope 2, and then forward into rope 3 |
| | So then here I am | 'Choo-choo' |
| | and this box gets smaller | Faster 'choo-choo,' about twice the |
| | when did I just get … Did I increase in size? | previous rate. Uses cupped hands to trace a circle centered on |
| | Kinetically? | her torso |
| | As a kinetic energy packet? | Arms down, hands slide laterally in and out |
| Andy | [Definitely] No. | Points to Sally |
| | So you're—so you're in here | Points to rope 3, steps back to rope 1 |
| Sally | So there's more of us in here? | |
| Andy | Yeah, yeah. Scuba Steve pushes the wall and so he's using his muscles, to do this | Pretends to be Scuba Steve and push wall inward |
| | and then we join you. | Drops arms suddenly, does 'choo-choo' as she jogs from rope 1 to rope 2 to rope 3 as energy unit |
| Sally | dah dah dah, dah dah dah | Sings along with Andy's motion |
| | So there's more packets of kinetic energy == | |
| Scott | | Steps into rope 3, rubs his palms together (group iconic gesture for thermal energy), changes to 'choo-choo' when Sally says 'kinetic energy' |
| Sally | == that's being used by the molecules. | |

In all three manners of enactment, energy is shown to increase: faster arms, larger body, and more persons. Is the progression toward representing more energy with more persons in this episode genuine intellectual progress or is it merely greater adherence to the arbitrary rules of the Energy Theater 'game'? For two reasons, we believe the progression of action is substantial intellectual development. First, only in the 'more persons' version of 'more energy' is the energy increase shown to result from the transfer of energy from one object to another. When arms move faster, or a person is imagined to grow larger, the energy might be shown to be increasing, but it is not shown to increase through a mechanism of flow or under a constraint of conservation. As we have shown previously (Scherr et al., 2013), the movement of persons to model the flow of energy leads naturally to learners considering the mechanisms and reasons for the flow of energy, which we believe is a more advanced level of analysis. In the present episode, the mechanism of energy transfer (the 'wall' or piston pushing on the molecules) is discussed verbally, but the link between the pushing and the energy flow are enacted explicitly only at the end, by Andy. Notice that Sally says 'It's giving it more kinetic energy to the molecules' after Scott explains the idea of energy flow verbally but appears not to appreciate the meaning of her own use of the word 'giving', that is, that energy flows from the wall, until Andy prepares to act it out by highlighting 'you're in here', and stepping back to rope 1. Thus it appears, in this situation at least, that the symbolic enactment of energy transfer through body motion is more effective at communicating the idea of flow of energy than verbal descriptions. One explanation in terms of blending is that Energy Theater, through its multi-modal nature, is the most efficient blended space for coordinating the many vital relations that concern us when explaining physical processes in terms of energy.

The second reason that we believe representing more energy with more persons is an intellectual advance for the group is that the transfer of logic from discrete domains to continuous domains has been shown to be generally more successful than the reverse (Bassok & Olseth, 1995). Setting aside the issue of whether energy is *actually* discrete or continuous, we recognize the importance in science of being able to model energy as either. If learners need to be able to model energy as continuous, it is better to practice modeling it as discrete, since the transfer from discrete to continuous is easier. Reasoning about continuous quantities has also been shown to be more successful when those quantities are parsed, or discretized (DeWolf, Bassok, & Holyoak, 2013). If 'more energy' were shown with more motion, or by imagining a bigger person, it would not be parsed, and so would not support the development of quantitative reasoning about energy.

Finally we discuss Andy's dual-role enactment first as Scuba Steve and then as an energy unit. Responding to Sally's 'Did I increase in size?', Andy moves over to rope 1, describes and copies Scuba Steve's actions, and then immediately changes her role to be a unit of kinetic energy that transfers from Steve, through the wall, and to the balls. First we notice that there are no objections by the group to Andy 'breaking character' as an energy unit to pretend to be an object in the scenario; nor is there any evidence of confusion as a result of Andy switching roles. On the contrary, Andy's actions are likely understood by the group as permissible, clear, and helpful; Sally, at least, appears to be satisfied with Andy's explanation. The situation poses a challenge to an analysis of the situation in terms of blending because of the apparent lack of consistency in the correspondence between the input spaces 'people' and 'scenario'. Our analysis is that Andy successfully and rapidly communicates the temporary engagement of a different identity (one for which she is blended with Scuba Steve) through her word choices and body action, and then similarly re-engages the Energy Theater blend by picking up the energy unit identity. Andy initiates the alternative blend between herself and Scuba Steve with the contrast in grammatical designation of Sally, using the second person 'You're in here', and herself, referring to herself as 'Scuba Steve' rather than 'I.' Through this choice of words, she indicates that she is something other than herself, or other than that with which her self has been blended thus far—namely, a unit of energy. When she is finished showing what Scuba Steve does, she becomes herself (that is, defined contextually, as blended with energy) again, which she indicates by referring to herself in the first person, saying 'we join you'. The change in role is also indicated with body action: As Scuba Steve, Andy holds her arms up as though she were pushing on large handles, imitating his body position, as visible in the simulation projected behind her. When she is finished showing what Steve does, her body position changes quickly—perhaps to communicate a discontinuous shift of identity—to begin a slow jog from rope 1 to rope 3, with her arms sustaining a continuous 'choo-choo' to communicate her identification with a unit of kinetic energy, while saying 'we join you'. Contrast this entire role-switching process with Dan's failed attempt to show an increase in pressure with double-high fives at the end of Video 1: Dan uses the first person 'we', formally proposes the action 'But what if we ... did this', and marks the proposal as symbolic by saying 'for pressure'. His bid is declined by the group because it is understood as an action proposed within the Energy Theater blend and with the same identity he had when he proposed the man-wall-balls pathway (rather than one in a blend running in parallel, in which he has an alternative identity) and as one that does not follow the Energy Theater blend's rules.

## Conclusions and Implications for Instruction

Our analysis of learner interactions during Energy Theater shows general agreement between the theory of conceptual blending and our observations. In particular, adult teachers-as-learners appear to find the blended space of Energy Theater to be usable, sensible, and productive. Further, the structure of the blended space appears to facilitate meaningful intellectual exchanges about the key features of energy dynamics, both in the gas compression scenario we studied here, and in other scenarios as we reported previously (Scherr et al., 2013). We note that in this study, in which we set out to analyze Energy Theater in terms of conceptual blending, we were compelled to analyze not only the planned, prescribed blend structure of Energy Theater, but also many improvised learner interactions that, in varying degrees, borrowed from, or varied from the prescribed blend, but which in any case had some significant blend structure. The intellectual work accomplished by learners in our study appears to have proceeded in large part through these improvised actions. We believe certain features of the design of Energy Theater promote these productive improvisations: rules that provide sufficient initial structure to the space and that generally promote multi-modal action, and, balancing the rule structure, plenty of opportunities for freely chosen action. Therefore, to those readers who would design similar activities for promoting science learning through blending and metaphor, we recommend that special attention be paid to the balance between rules and free choice that is characteristic of structured improvisation.

We aim to promote Energy Theater as a valuable classroom learning activity for teachers and students from young adolescence to adulthood in all science disciplines. Our hypothesis is that teachers and students that use Energy Theater will more reliably conserve energy as they track its transfers and transformations among objects in a system, and that Energy Theater does so through its leveraging of some fundamental cognitive patterns of conceptual blending.

Our work demonstrates the instructional effectiveness of Energy Theater with a small number of adult learners. Further research is needed to examine wider implications for practice. For example, future investigations might use methodologies similar to ours to study *student* learning as a result of Energy Theater, or use larger-scale methodologies (such as survey instruments or written assessments) to study the learning of larger numbers of teachers and their students. We also hope to conduct studies of what scaffolding is needed for younger learners to engage meaningfully with Energy Theater. Such investigations will inform the development of classroom practices that greatly enhance the learning of energy conservation and tracking.

## Disclosure statement

No potential conflict of interest was reported by the authors.

## Funding

This work was supposed by the National Science Foundation [grant number DRL 0822342].

## Supplemental data

Supplemental data for this article can be accessed at http://dx.doi.org/10.1080/09500693.2015.1025307.

## Notes

1. Some of this material appeared previously in Scherr et al. (2012).
2. See Amin (2009) for more examples of the manner in which Feynman speaks metaphorically about energy.

## References

Amin, T. G. (2009). Conceptual metaphor meets conceptual change. *Human Development, 52*(3), 165–197.

Anderson-Levitt, K. M. (2006). Ethnography. In J. L. Green, G. Camilli, P. B. Elmore, A. Skukaus-kaitė, & E. Grace (Eds.), *Handbook of complementary methods in education research* (pp. 279–295). Mahwah, NJ: Lawrence Erlbaum Associates.

Arons, A. B. (1965). *Development of concepts of physics: From the rationalization of mechanics to the first theory of atomic structure.* Reading, MA: Addison-Wesley.

Bassok, M., & Olseth, K. L. (1995). Object based representations: Transfer between cases of continuous and discrete models of change. *Journal of Experimental Psychology: Learning, Memory, and Cognition, 21*(6), 1522–1538.

Begel, A., Garcia, D. D., & Wolfman, S. (2004, March 3). *Kinesthetic learning in the classroom.* Paper presented at the proceedings of the 35th SIGCSE technical symposium on computer science education SIGCSE '04, New York, NY.

Brewe, E. (2011). Energy as a substance like quantity that flows: Theoretical considerations and pedagogical consequences. *Physical Review Special Topics: Physics Education Research, 7*(020106), 1–14.

Chi, M. T. H. (2005). Commonsense conceptions of emergent processes: Why some misconceptions are robust. *Journal of the Learning Sciences, 14*(2), 161–199.

Chi, M. T. H., & Slotta, J. D. (1993). The ontological coherence of intuitive physics. *Cognition and Instruction, 10*(2–3), 249–260.

Chinnicci, J. P., Yue, J. W., & Torres, K. M. (2004). Students as 'human chromosomes' in roleplaying mitosis and meiosis. *The American Biology Teacher, 66*(1), 35–39.

Close, H. G., & Scherr, R. E. (2012). Differentiation of energy concepts through speech and gesture in interaction. *American Institute of Physics Conference Proceedings, 1413*, 151–154.

Colella, V. (2000). Participatory simulations: Building collaborative understanding through immersive dynamic modeling. *Journal of the Learning Sciences, 9*(4), 471–500.

Daane, A. R., McKagan, S. B., Vokos, S., & Scherr, R. E. (2015). Energy conservation in dissipative processes: Teacher expectations and strategies associated with imperceptible thermal energy. *Physical Review – Special Topics: Physics Education Research, 11*(010109), 1–15.

Daane, A. R., Vokos, S., & Scherr, R. E. (2014). Goals for teacher learning about energy degradation. *Physical Review Special Topics: Physics Education Research, 10*(020111), 1–16.

Daane, A. R., Wells, L., & Scherr, R. E. (2014). Energy theater. *The Physics Teacher, 52*, 291–294.

Denzin, N. K., & Lincoln, Y. S. (2005). Introduction: The discipline and practice of qualitative research. In N. K. Denzin & Y. S. Lincoln (Eds.), *The Sage handbook of qualitative research* (4th ed., pp. 1–19). Thousand Oaks, CA: Sage Publications.

DeWolf, M., Bassok, M., & Holyoak, K. J. (2013). *Analogical reasoning with rational numbers: Semantic alignment based on discrete versus continuous quantities.* Paper presented at the 35th Annual Conference of the Cognitive Science Society, Austin, TX.

diSessa, A. (1993). Ontologies in pieces: Response to Chi and Slotta. *Cognition and Instruction, 10*(2–3), 272–280.

Dreyfus, B. W., Geller, B. D., Gouvea, J., Sawtelle, V., Turpen, C., & Redish, E. F. (2014). Ontological metaphors for negative energy in an interdisciplinary context. *Physical Review Special Topics: Physics Education Research, 10*(020108), 1–11.

Duit, R. (1987). Should energy be illustrated as something quasi material? *International Journal of Science Education, 9*(2), 139–145.

Erickson, F. (1986). Qualitative methods in research on teaching. In M. C. Wittrock (Ed.), *Handbook of research on teaching* (pp. 119–161). New York, NY: Macmillan.

Erickson, F. (2004). *Talk and social theory: Ecologies of speaking and listening in everyday life.* Malden, MA: Polity Press.

Falk, G., Hermann, F., & Bruno Schmid, G. (1983). Energy forms or energy carriers? *American Journal of Physics, 51*(12), 1074–1077.

Fauconnier, G., & Turner, M. (2002). *The way we think: Conceptual blending and the mind's hidden complexities.* New York, NY: Basic Books.

Feynman, R. P., Leighton, R. B., & Sands, M. (1969). *The Feynman lectures on physics.* Reading, MA: Addison Wesley.

Goodwin, C. (1994). Professional vision. *American Anthropologist, 96*(3), 606–633.

Goodwin, C. (2000). Action and embodiment within situated human interaction. *Journal of Pragmatics, 32*(10), 1489–1522.

Greeno, J. G. (1998). The situativity of knowing, learning, and research. *American Psychologist*, *53*(1), 5. doi:10.1037/0003-066X.53.1.5

Halliday, D., Resnick, R., & Walker, J. (2008). *Fundamentals of physics* (8th ed.). Hoboken, NJ: John Wiley & Sons.

Hutchins, E. (1995). How a cockpit remembers its speeds. *Cognitive Science*, *19*(3), 265–288.

Hutchins, E. (2005). Material anchors for conceptual blends. *Journal of Pragmatics*, *37*(10), 1555–1577.

Jordan, B., & Henderson, A. (1995). Interaction analysis: Foundations and practice. *Journal of the Learning Sciences*, *4*(1), 39–103.

Kraus, P. A., & Vokos, S. (2011). The role of language in the teaching of energy: The case of heat energy. *Washington State Teachers' Association Journal*, Spring. Retrieved from http://www.spu.edu/depts/physics/documents/wsta_krausvokos.pdf

Lakoff, G. (1987). *Women, fire, and dangerous things: What categories reveal about the mind*. Chicago, IL: University of Chicago Press.

Lakoff, G., & Johnson, M. (1999). *Philosophy in the flesh: The embodied mind and its challenge to Western thought*. New York, NY: Basic Books.

Lakoff, G., & Nuñez, R. E. (2000). *Where mathematics comes from: How the embodied mind brings mathematics into being*. New York, NY: Basic Books.

Lave, J. (1991). Situating learning in communities of practice. In L. B. Resnick, J. M. Levine, & S. D. Teasley (Eds.), *Perspectives on socially shared cognition* (pp. 63–82). Washington, DC: American Psychological Association.

Manogue, C. A., Siemens, P. J., Tate, J., Browne, K., Niess, M. L., & Wolfer, A. J. (2001). Paradigms in physics: A new upper-division curriculum. *American Journal of Physics*, *69*(9), 978–990.

Maxwell, J. A. (2004a). Causal explanation, qualitative research, and scientific inquiry in education. *Educational Researcher*, *33*(2), 3–11.

Maxwell, J. A. (2004b). Using qualitative methods for causal explanation. *Field Methods*, *16*(3), 243–264.

Mcdermott, R. P., Gospodinoff, K., & Aron, J. (1978). Criteria for an ethnographically adequate description of concerted activities and their contexts. *Semiotica*, *24*(3–4), 245–275.

McKagan, S. B., Scherr, R. E., Close, E. W., & Close, H. G. (2012). Criteria for creating and categorizing forms of energy. *American Institute of Physics Conference Proceedings*, *1413*, 279–282.

Millar, R. (2005). *Teaching about energy*. York: Department of Educational Studies Research Paper, York University.

Morrow, C. A. (2000). Kinesthetic astronomy: The sky time lesson. *The Physics Teacher*, *38*, 252–253. doi:10.1119/1.880520

National Research Council. (2012). *A framework for K-12 science education: Practices, crosscutting concepts, and core ideas*. Washington, DC: National Academies Press.

Nemirovsky, R., Rasmussen, C., Sweeney, G., & Wawro, M. (2011). When the classroom floor becomes the complex plane: Addition and multiplication as ways of bodily navigation. *Journal of the Learning Sciences*, *21*(2), 287–323.

NGSS Lead States. (2013). *Next Generation Science Standards: For states, by states*. Washington, DC: National Academies Press.

Ochs, E., Gonzales, P., & Jacoby, S. (1996). 'When I come down I'm in the domain state': Grammar and graphic representation in the interpretive activity of physicists. In E. Ochs, E. Schegloff, & S. Thompson (Eds.), *Interaction and grammar* (pp. 328–369). Cambridge: Cambridge University Press.

Reinfeld, E. L., & Hartman, M. A. (2008). Kinesthetic life cycle of stars. *Astronomy Education Review*, *7*(2), 158–175.

Resnick, M., & Wilensky, U. (1998). Diving into complexity: Developing probabilistic decentralized thinking through role-playing activities. *Journal of the Learning Sciences*, *7*(2), 153–172.

Richards, T. (2010). Using kinesthetic activities to teach Ptolemaic and Copernican retrograde motion. *Science and Education*, *21*(6), 899–910.

Ross, P. M., Tronson, D. A., & Ritchie, R. J. (2008). Increasing conceptual understanding of glycolysis and the Krebs cycle using role play. *The American Biology Teacher, 70*(3), 163–168.

Salmon, W. C. (1998). *Causality and explanation*. New York, NY: Oxford University Press.

Schegloff, E. (1997). Whose text? Whose context? *Discourse Society, 8*(2), 165–187.

Scherr, R. E., Close, H. G., Close, E. W., Flood, V. J., McKagan, S. B., Robertson, A. D., & Vokos, S. (2013). Negotiating energy dynamics through embodied action in a materially structured environment. *Physical Review Special Topics: Physics Education Research, 9*(020105), 1–18.

Scherr, R. E., Close, H. G., Close, E. W., & Vokos, S. (2012). Representing energy. II. Energy tracking representations. *Physical Review Special Topics: Physics Education Research, 8*(020115), 1–11.

Scherr, R. E., Close, H. G., McKagan, S. B., & Vokos, S. (2012). Representing energy. I. Representing a substance ontology for energy. *Physical Review Special Topics: Physics Education Research, 8*(020114), 1–11.

Scherr, R. E., & Robertson, A. D. (2014). The productivity of 'collisions generate heat' for making sense of energy transformations in adiabatic compression: A case study. *Physical Review – Special Topics: Physics Education Research, 11*(010111), 1–16. Retrieved from http://dx.doi.org/10.1103/PhysRevSTPER.11.010111

Serway, R. A., & Jewett, J. W. (2007). *Physics for scientists and engineers with modern physics* (7th ed.). Belmont, CA: Thomson Higher Education.

Sfard, A. (1994). Reification as the birth of metaphor. *For the Learning of Mathematics, 14*(1), 44–55.

Sfard, A. (1998). On two metaphors for learning and the dangers of choosing just one. *Educational Researcher, 27*(2), 4–13.

Sfard, A. (2007). When the rules of discourse change, but nobody tells you: Making sense of mathematics learning from a commognitive standpoint. *Journal of the Learning Sciences, 16*(4), 565–613.

Sherin, B. L. (2001). A comparison of programming languages and algebraic notation as expressive languages for physics. *International journal of Computers for Mathematics Learning, 6*(1), 1–61.

Singh, V. (2010). The electron runaround: Understanding electric circuit basics through a classroom activity. *The Physics Teacher, 48*(5), 309–311.

Slotta, J. D., & Chi, M. T. H. (2006). Helping students understand challenging topics in science through ontology training. *Cognition and Instruction, 24*(2), 261–289.

Stevens, R. (2000). Divisions of labor in school and in the workplace: Comparing computer and paper supported activities across settings. *Journal of the Learning Sciences, 9*(4), 373–401.

Stevens, R. (2012). The missing bodies of mathematical thinking and learning have been found. *Journal of the Learning Sciences, 21*(2), 337–346.

Swackhamer, G. (2005). *Cognitive resources for understanding energy*. Retrieved from http://modeling.la.asu.edu/modeling/CognitiveResources-Energy.pdf

Touval, A., & Westreich, G. (2003). Teaching sums of angle measures: A kinesthetic approach. *The Mathematics Teacher, 96*(4), 230–237.

Vygotsky, L. S. (1986). The development of scientific concepts in childhood: The design of a working hypothesis. In A. Kozulin (Ed.), *Thought and language* (pp. 146–209). Cambridge, MA: MIT Press.

Warren, J. (1982). The nature of energy. *European Journal of Science Education, 4*, 295–297.

Warren, J. (1986). At what stage should energy be taught? *Physics Education, 21*, 154–156.

Weiman, C. E., Adams, W. K., & Perkins, K. K. (2008). PHYSICS: PhET: Simulations that enhance learning. *Science, 322*(5902), 682–683.

Wertsch, J. V. (2007). Mediation. In H. Daniels, M. Cole, & J. V. Wertsch (Eds.), *The Cambridge companion to Vygotsky* (pp. 178–192). New York, NY: Cambridge University Press.

Wyn, M. A., & Stegnik, S. J. (2000). Role playing mitosis. *The American Biology Teacher, 62*(5), 378–381.

Zimmerman, V. (2002). Moving poems: Kinesthetic learning in the literature classroom. *Pedagogy: Critical Approaches to Teaching Literature, Language, Composition, and Culture, 2*(3), 409–412.

# Some Challenges in the Empirical Investigation of Conceptual Mappings and Embodied Cognition in Science Education: Commentary on Dreyfus, Gupta and Redish; and Close and Scherr

Rafael Núñez

*Department of Cognitive Science, University of California, San Diego, CA, USA*

The last couple of decades have seen an enormous development in the study of embodied cognition through the investigation of conceptual mappings, such as conceptual metaphor (Lakoff & Johnson, 1980/2003) and conceptual blending (Fauconnier & Turner, 2002). Initially, this progress was achieved at a theoretical level, and more recently through empirical research in basic science—from psycholinguistics, to cross-cultural and developmental studies, to cognitive neuroscience (for a collection of review chapters, see, for example, Gibbs, 2008; see also, Fauconnier & Turner, 2002; Lakoff & Núñez, 2000). These advancements have begun to be applied to domains such as literary criticism (Turner, 1998), advertising (Joy, Sherry, & Deschenes, 2009), law and courtroom settings (Pascual, 2008), theater (Cook, 2007), and, importantly, they have reached the critical sphere of education in mathematics (e.g. Edwards, 2009; Zandieh, Roh, & Knapp, 2014) and science (Hrepic, Zollman, & Rebello, 2010). The present issue constitutes an excellent example of how science education is reaching new levels of research maturity and sophistication, bringing advances from contemporary cognitive science to the study of the richness and complexity involved in the teaching and learning of science—a laudable endeavor. Beyond the already multilayered practice of teaching and implementing educational curricula, taking the step of conducting empirical research in the domain of embodied cognition and conceptual mappings in educational settings brings new challenges and difficulties. Here I will analyze some of these difficulties, hoping that they make a contribution to the design and implementation of future empirical research in these domains. Although I will refer specifically to two articles of this special issue (Dreyfus, Gupta, & Redish, 2015; Close & Scherr, 2015), the discussion should be generic enough so that it can be applied to other research settings that investigate embodied cognition, conceptual mappings and gesture in mathematics and science education.

The papers by Dreyfus et al., and by Close and Scherr, share several features. Both articles investigate the relationship between embodied cognition, conceptual mappings, and gesture production in the domain of energy as studied in college physics. And, methodologically, both present studies conducted in the descriptive and interpretive tradition, and both choose to focus their analysis on two episodes carefully selected from a remarkably large video-recorded database. These papers analyze in detail the ENERGY AS A SUBSTANCE[1] conceptual metaphor, and its relation to other conceptual mappings, gesture production, and notation systems. Dreyfus et al. study how the inferential organization of the ENERGY AS A SUBSTANCE metaphor might interact with a conceptual metaphor they call ENERGY AS VERTICAL LOCATION, both being orchestrated via an encompassing conceptual blend. Close and Scherr, working with teachers involved in a professional development program, study the ENERGY AS A SUBSTANCE metaphor in the context of an 'Energy Theater', a pedagogical setting that builds on a role-playing situation in which teacher–learners/actors explore and enact the role of various relevant entities involving energy transfers and transformations. Both these articles do a wonderful job of describing the fundamentals of conceptual metaphor and conceptual blending, and I applaud their efforts in describing, explaining, and engaging with, these subtle and sophisticated theories, applying them to specific contexts of physics education. Indeed, these papers provide excellent examples of how complex multi-modal processes involving abstract technical conceptual systems, language, communication and bodily actions unfold in real time, enacting dynamic sense-making in real-world scenarios. The empirical investigation of such complex phenomena is, of course, far from easy, as they involve a great deal of methodological and theoretical problems when it comes to designing studies, operationalizing relevant variables, and gathering, analyzing, and interpreting data. In the spirit of constructive criticism, I raise here a few points that are problematic in these papers, which usually present difficulties when investigating embodied cognition, conceptual mappings, and gesture production in general.

## What is in a Domain or Input Space of a Conceptual Mapping?

Traditionally, conceptual metaphor and blending theories would postulate the existence of source and target domains of conceptual metaphors or input spaces, respectively, based on linguistic data—that is, actual figurative linguistic expressions. For instance, metaphorical expressions, such as *this theory has weak foundations*, would suggest that their semantics draw from the inferential organization provided by a systematic mapping from entities in a source domain of Buildings onto elements in the target domain of Theories, hence the name of the conceptual metaphor THEORIES ARE BUILDINGS (Lakoff & Johnson, 1980/2003). Researchers in cognitive linguistics were quick to point out, however, that many elements of the domain of Buildings do not get mapped onto Theories: restrooms, elevators, and so on (Grady, 1997). Thus, an important area of research in conceptual mappings became the precise character-ization of what exactly constitutes a domain (or input space) structuring the inferen-tial organization of the mapping, and on what bases should the researcher decide what the domain (or input) spaces are supposed to be. This problem comes up, in Dreyfus et al.'s paper. The authors want to show how a 'single blended mental space' (p. 1) blends two ontological metaphors: ENERGY AS A SUBSTANCE and ENERGY AS A VER-TICAL LOCATION. Although both conceptual metaphors share the same target domain of ENERGY, one wonders what exactly constitutes the source domain of the latter: VERTICAL LOCATION. A quick look at the authors' reported video data reveals that this source domain is not just about the ontology of a generic ordinary 'vertical location' as in a person's height or the elevation of a hill, but that it largely (if not entirely) relies on the technical conceptual system (and notational apparatus) of the Cartesian Plane and Analytic Geometry, with specific variables mapped on the $x$- and $y$-axes (distance between atoms and energy, respectively). The blended space that the authors are trying to analyze is therefore, not between two ontological every-day-like conceptual metaphors involving a generic ordinary notion of 'substance' and of 'vertical location', but in the case of the latter, one that involves a highly technical domain mediated by notation, graphic conventions, and further abstracted concepts. 'Vertical location' here is not just a location in space standing apart from the bodily experienced ground, but a specific construal based on $y$-coordinates of points depicted graphically on an external medium such as a classroom board. This distinc-tion is not purely formal, as it points to the difference between the ontology of meta-phorical source domains based on ordinary everyday bodily experience (e.g. people's height) and technical domains that are mediated by specific conceptual systems (e.g. Analytic Geometry and the Cartesian Plane) constrained and regulated by highly con-ventionalized norms, notations, and inscriptions. These two types of domains have quite different inferential organization and range of applicability. In fact, any domain—not just energy—that is susceptible to be analyzed in terms of graphically depicted functions in the Cartesian Plane—stock markets, number of infected people in pandemics, rainfalls, cholesterol levels, and so on, would essentially support similar forms of reasoning and sense-making in terms of 'ups' and 'downs'. Thus, the statement from a physics professor—Prof. Farnsworth—quoted in Dreyfus et al.'s paper (p. 16) in which, he utters, as he refers to a graph depicted on the board, '[the energy levels of two atoms] *drop down to here*', does not primarily express energy concepts in terms of a source domain of bodily grounded (vertical) space, but rather in terms of a conceptual algebraic-geometric technicalized domain determined by the Cartesian Plane, depicted in an external medium. Importantly, it is this distinction that allows us to understand (1) what is the role of certain types of gestures in conceptual blending, and (2) why the putative metaphor ENERGY AS

VERTICAL LOCATION elicits such an abundant amount of utterances that involve demonstratives (e.g. 'this' and 'that'), deictic terms (e.g. 'here' and 'there'), and specific types of indexical gestures (i.e. pointings). Let us analyze these observations in the next section.

## Gestures as Evidence of Mappings? Or as Constituting Input Spaces? Or as Doing Something Else?

Both the article by Dreyfus et al., and the one by Close and Scherr describe interesting instances of gestures co-produced with rich conceptual elaborations. Close and Scherr analyze gestures in which the teachers–learners/actors in the 'Energy Theater' take on the role of energy units in a problem scenario. Not only the theater context calls for a conceptual blend that involves a degree of personification (i.e. teachers–learners 'being' energy units; see 'Drama connectors' in Fauconnier & Turner, 2002), but, interestingly, it also prompts occasional opportunistic impersonations outside of the rules of the game, which must be managed at a meta-discursive/conceptual level. During brief passages, teachers–learners *are not* units of energy, but some other invoked characters or entities. Thus, we see teacher–learner Andy spontaneously impersonating a scuba diver who pushes the walls of a container with a gas in it, and teacher–learner Sally making a machine-like gesture—a simulation of 'the pumping action of locomotive wheels with bent arms pumping forward and backward close to the body' (p. 32), which is a gesture with iconic content that participants have agreed to use to refer to 'kinetic energy'. In the gesture studies literature, these types of gesture are called 'pantomimes' (McNeill, 2000), which are characterized by having a first person perspective, by extensively using parts of the body (or all of it), and by displaying macro movements in space that portray some prominent features of the entities or characters being invoked. When pantomimes—spontaneous and conventionalized—are produced in the context of theater scenarios, it is relatively straightforward to follow arguments that explain their enactment in terms of blended spaces—with an input space constituted by the actual individual who produces bodily actions in his/her surrounding space (e.g. the teachers–learners Andy and Sally), and an input space constituted by the character being impersonated (e.g. the scuba diver, and a locomotive displaying the pumping action of its wheels, respectively). It is in the blended space that one can interpret the bodily actions of actors as being those from the character being played (Cook, 2007; Fauconnier & Turner, 2002). Thus, in Close and Scherr's paper, for instance, we unproblematically read (p. 32): With 'embodied actions' Andy (a teacher–learner) 'pretends to be Scuba Steve and push wall inward'.

The gestures described by Dreyfus et al., however, present some challenges. The authors describe gestures produced by the physics professor, Farnsworth, while teaching, and those produced during an interview by an undergraduate pre-medical student taking his class. And they explicitly declare that they use 'gestures as evidence for an underlying conceptual blend rather than considering the gestures themselves as an input to the blend' (p. 11). Certain types of gestures do in fact contribute to providing evidence of an underlying conceptual blend. Such is the case of the pantomimes described above, as well as that of spontaneous gestures that bring semantic structure that is not expressed in the speech modality (Cienki, 1998), for instance when someone utters 'in my childhood' while pointing backwards. In such case, 'childhood' is brought as a temporal entity that, in itself, does not have spatial (metaphorical) content, and the backwards pointing is a motor action that indexes a physical space behind the speaker. It is the blended ensemble that reveals the specificities of the spatial construal of the temporal entity 'childhood' as metaphorically located behind the speaker. But most of the gestures reported and described by Dreyfus et al. are of a different nature and demand extra caution in the analysis. Many of these gestures are pointings that co-occur with demonstratives and deictic terms, co-produced with utterances such as 'and drop down to *here*', 'and release *that* much energy', 'and *that* negative energy', 'come in at *this* energy', 'you gotta put in *this* much', 'the difference between *here* and *here*', and so on (pp. 17–21; italics added). To state that these gestures provide evidence for an underlying conceptual blend of ontological metaphors 'where the predicate from one ontological category is accompanied by gestures from another' (p. 19) is misleading and problematic. To illustrate the special status of pointings co-produced with demonstratives (or deictic terms), consider a case in which someone utters 'I prefer this one over that one'. In the pure speech/linguistic track, the utterance is completely underspecified such that almost no coherent meaning can be enacted. Collections of utterances of this sort, therefore, cannot constitute a (source or input) domain for a conceptual mapping, let alone an ontological domain. Indeed, in everyday communication, utterances of this sort *never* occur without some type of co-produced bodily action: hand or tool pointings, eye-gaze, etc. And if, for some reason, they do occur (e.g. because of inattention on the part of the speaker) they are remedied right away in the communicative process. As we saw in the previous section, the content involving the putative ENERGY AS A VERTICAL LOCATION in Dreyfus et al.'s data is in fact specified by the Cartesian Plane and mediated by the graphs and inscriptions that are externalized on the board. The demonstratives (this and that) and deictic terms (here and there) that appear in the speech modality actually index specific technical entities depicted or written on the board, and therefore, the observed co-occurring pointings are not manifestations of gestures from an ontological domain to be blended. Rather, they are specific types of gestures whose function is to make the indexation process unambiguous. These indexical gestures (but not their morphologies), therefore, are entirely inscription- and notation-dependent, and therefore they cannot be taken as evidence of a conceptual blending of ontological metaphors. They need to be handled with care when it comes to analyze them in terms of embodied cognition and conceptual mappings.

## Closing Remarks on Methods, Research Issues, and Theory Building

Historically, conceptual metaphor theory, which initially developed on the basis of linguistic and theoretical analysis, ended up benefiting from criticism from the empirical

127

sciences for its ultimate development. Psychologists, early in the process, argued that if the theory was supposed to be not just about 'verbal' or 'linguistic' metaphors, but about concepts, thought, and reasoning, then specific evidence had to be provided to claim that when, say, a woman tells her partner *we have been walking through life together* she is *actually* cognitively operating with the LOVE IS A JOURNEY metaphor (Murphy, 1996). In the late nineties, scores of experiments began to be conducted to test, beyond purely linguistic expressions and etymologies, the 'psychological reality' of many of the postulated conceptual metaphors. Initially, these studies involved carefully controlled psycholinguistic experiments in the laboratory (e.g. Gentner & Wolff, 1997), and later they were extended to fieldwork and cross-cultural settings involving other methods such as gesture analysis (e.g. Núñez & Sweetser, 2006). As a result, a deeper and more detailed understanding of metaphorical thinking was fostered. In most of these studies, researchers exploited the fact that metaphorical mappings are essentially unidirectional—they go from a source domain to a target domain—making experimentation with conceptual metaphors empirically manageable. For instance, using priming studies, researchers could experimentally manipulate (i.e. 'prime') the source domain of a spatio-temporal mapping (e.g. by exposing participants to certain types of *spatial* material), and proceed to empirically test predictions about what *temporal* inferences in the target domain participants would make as a result of the manipulation (e.g. Núñez, Motz, & Teuscher, 2006). With respect to conceptual blending, however, the experimental investigation of the mappings has been much more difficult to conceive, to design, and to conduct. The underlying problems are largely due to the fact that, unlike conceptual metaphor, blending mappings are essentially non-directional, and the postulated networks of mappings are often complex and hard to deal with operationally. To partially remedy the situation, some improvements have come from gesture studies (e.g. Parrill & Sweetser, 2004), which have broadened the range of available empirical methods. But there is still much more to be done to put conceptual blending theory (and conceptual mappings in general) on firm empirical grounds.

Considering the difficulties that cognitive linguists and cognitive scientists have encountered in empirically testing hypotheses involving conceptual mappings, one can only praise the efforts deployed by Dreyfus et al. and Close and Scherr in their insightful and perceptive studies in science education. In order to make progress in the future, however, it would be highly advisable if more efforts were put to (1) move from purely *interpretive* studies to more methodologically rigorous ones, (2) move from a mainly *confirmatory* approach designed to corroborate pre-existing beliefs, to an approach that attempts also to *disconfirm* alternative explanatory proposals, and, ultimately, (3) define more precise standards for what count as *evidence* in research. We read in Dreyfus et al., for example, that although 'the episodes selected for analysis are intended to be illustrative of what ontological blending looks like' (p. 12), they nonetheless reach the conclusion that 'this analysis yields evidence that speakers are blending the substance and location ontologies into a single mental space' (p. 1). But, is there really evidence of that? Similarly, based also on a purely interpretive method, Close and Scherr affirm that they 'demonstrate that a particular blended learning space is especially productive in developing understanding of energy transfers and transformations' (p. 1). While the authors' assertions may not be incorrect, it is not clear, on empirical grounds, that their interpretations constitute evidence proper, or that they hold because of the reasons (or explanations) the authors provide. The 'Energy Theater' scenario of Close and Scherr, for instance, may be productive not because of the 'acting' blend of teachers–learners, but because of underlying attentional and motivational factors that could operate also on other, non-first person-driven blends, such as external impersonations using wooden toys. These may be questions that originate theoretically or from extended professional experience with teaching, but ultimately, they are empirical questions that could be answered by carefully designing studies that attempt to explain phenomena not only by confirming cases, but crucially, by excluding or *disconfirming* alternative explanatory possibilities. From this perspective, 'evidence' may not necessarily be constituted by a cherry-picked example that confirms the researchers' beliefs, but by an exhaustive analysis of cases through which alternative hypotheses, explanations, and interpretations are excluded. The papers analyzed here present two pairs of carefully picked episodes taken from large video databases. It would therefore be desirable that the analyses are conducted not just on two ad-hoc examples that confirm or illustrate the authors' opinions and impressions, but on a larger collection of episodes that might shed light on alternative explanatory proposals. Dreyfus et al., for example, overtly write 'our analysis of gestures is interpretive rather than following a systematic coding scheme … In lieu of that, we provide enough details in the data analysis so that readers can draw their conclusions and evaluate our gesture analysis' (p. 14). While the authors might be right that a systematic coding scheme may be unnecessary for their purposes, it is not the case that the readers can freely draw conclusions that might interpret the data differently, because they only have access to the two transcribed episodes picked by the authors and not to the rich video-recorded database where they might find alternative explanations to the reported phenomena. It is healthy for young fields of investigation, such as embodied cognition and conceptual mappings, to pass through an extended period describing phenomena in detail. Dreyfus et al.'s and Close and Scherr's papers give us excellent examples of subtle, perceptive, and insightful studies that investigate conceptual mappings, embodied cognition and gestures in the context of science education. But it is also important to consider the future of research in these areas. For that, in order to achieve maturity with respect to explanatory power and theory development, a field of research must go beyond purely descriptive approaches, and gradually incorporate empirical research methods that, seeking for causes and explanations, rigorously constrain the universe of potential interpretations (Núñez, 2012). The present special issue provides wonderful examples of nuanced and thoughtful high-quality descriptive studies in embodied cognition and conceptual mappings in science education. Now the time seems ripe for taking the empirical research in this domain to exciting new territories, using the very scientific method constitutive of the subject matter of *science* education.

## Disclosure statement

No potential conflict of interest was reported by the author.

## Note

1. Following a convention in cognitive linguistics, here I denote the name of a conceptual metaphor in small capitals, as in AFFECTION IS WARMTH, so it can be distinguished from specific linguistic instantiations—metaphorical expressions—such as *send her my warm helloes*, which I will denote in italics.

## References

Cienki, A. (1998). Metaphoric gestures and some of their relations to verbal metaphoric counterparts. In J.-P. Koenig (Ed.), *Discourse and cognition: Bridging the gap* (pp. 189–205). Stanford, CA: CSLI.

Close, H. G., & Scherr, R. E. (2015). Enacting conceptual metaphor through blending: Learning activities embodying the substance metaphor for energy. *International Journal of Science Education.* doi:10.1080/09500693.2015.1025307

Cook, A. (2007). Interplay: The method and potential of a cognitive scientific approach to theatre. *Theatre Journal, 59*(4), 579–594.

Dreyfus, B. W., Gupta, A., & Redish, E. F. (2015). Applying conceptual blending to model coordinated use of multiple ontological metaphors. *International Journal of Science Education.* doi:10.1080/09500693.2015.1025306

Edwards, L. D. (2009). Gestures and conceptual integration in mathematical talk. *Educational Studies in Mathematics, 70*(2), 127–141.

Fauconnier, G., & Turner, M. (2002). *The way we think: Conceptual blending and the mind's hidden complexities.* New York: Basic Books.

Gentner, D., & Wolff, P. (1997). Alignment in the processing of metaphor. *Journal of Memory and Language, 37*(3), 331–355.

Gibbs Jr, R. W. (Ed.). (2008). *The Cambridge handbook of metaphor and thought.* Cambridge: Cambridge University Press.

Grady, J. E. (1997). Theories are buildings revisited. *Cognitive Linguistics (includes Cognitive Linguistic Bibliography), 8*(4), 267–290.

Hrepic, Z., Zollman, D. A., & Rebello, N. S. (2010). Identifying students' mental models of sound propagation: The role of conceptual blending in understanding conceptual change. *Physical Review Special Topics-Physics Education Research, 6*(2), 020114.

Joy, A., Sherry, J. F., & Deschenes, J. (2009). Conceptual blending in advertising. *Journal of business research, 62*(1), 39–49.

Lakoff, G., & Johnson, M. (1980/2003). *Metaphors we live by.* Chicago: University of Chicago Press.

Lakoff, G., & Núñez, R. E. (2000). *Where mathematics comes from: How the embodied mind brings mathematics into being.* New York: Basic books.

McNeill, D. (Ed.). (2000). *Language and gesture* (Vol. 2). Cambridge: Cambridge University Press.

Murphy, G. L. (1996). On metaphoric representation. *Cognition, 60*(2), 173–204.

Núñez, R. (2012). On the science of embodied cognition in the 2010s: Research questions, appropriate reductionism, and testable explanations. *Journal of the Learning Sciences, 21*(2), 324–336.

Núñez, R. E., Motz, B. A., & Teuscher, U. (2006). Time after time: The psychological reality of the ego-and time-reference-point distinction in metaphorical construals of time. *Metaphor and Symbol, 21*(3), 133–146.

Núñez, R. E., & Sweetser, E. (2006). With the future behind them: Convergent evidence from Aymara language and gesture in the crosslinguistic comparison of spatial construals of time. *Cognitive Science, 30*(3), 401–450.

Parrill, F., & Sweetser, E. (2004). What we mean by meaning: Conceptual integration in gesture analysis and transcription. *Gesture, 4*(2), 197–219.

Pascual, E. (2008). Fictive interaction blends in everyday life and courtroom settings. In T. Oakley & A. Hougaard (Eds.), *Mental spaces in discourse and interaction* (pp. 79–108). Amsterdam: John Benjamins Publishing Company.

Turner, M. (1998). *The literary mind: The origins of thought and language.* Oxford: Oxford University Press.

Zandieh, M., Roh, K. H., & Knapp, J. (2014). Conceptual blending: Student reasoning when proving "conditional implies conditional" statements. *The Journal of Mathematical Behavior, 33*(March), 209–229.

# An Analysis of Metaphors Used by Students to Describe Energy in an Interdisciplinary General Science Course

Rachael Lancor

*Department of Chemistry, Geoscience and Physics, Edgewood College, Madison, WI , USA*

The meaning of the term *energy* varies widely in scientific and colloquial discourse. Teasing apart the different connotations of the term can be especially challenging for non-science majors. In this study, undergraduate students taking an interdisciplinary, general science course ($n = 49$) were asked to explain the role of energy in five contexts: radiation, transportation, generating electricity, earthquakes, and the big bang theory. The responses were qualitatively analyzed under the framework of conceptual metaphor theory. This study presents evidence that non-science major students spontaneously use metaphorical language that is consistent with the conceptual metaphors of energy previously identified in the discourse of students in introductory physics, biology, and chemistry courses. Furthermore, most students used multiple coherent metaphors to explain the role of energy in these complex topics. This demonstrates that these conceptual metaphors for energy have broader applicability than just traditional scientific contexts. Implications for this work as a formative assessment tool in instruction will also be discussed.

## Introduction

The exact meaning of the term energy depends on disciplinary context. Anecdotally, many teachers have noticed that students compartmentalize disciplinary ideas about energy; students think energy in biology is different from energy in physics. One way to understand these different conceptualizations of energy is through the lens of conceptual metaphor theory (Amin, 2009; Dreyfus et al., 2014; Lancor, 2014a; Scherr, Close, McKagan, & Vokos, 2012). Conceptual metaphor theory is a cognitive theory that argues that the way we understand the world is largely metaphoric in nature (Lakoff & Johnson, 1980, 1999). Metaphorical language is necessary to articulate and comprehend abstract ideas, and conceptual metaphor theory provides a way for researchers to gain insight into how students understand abstract concepts, such as energy. Introductory physics classes may primarily use energy as an accounting system to track changes in a system, while energy in an introductory biology class may primarily be presented as a substance that can be lost from a system. Although conscientious students may see connections between how energy is used in different contexts, most students are not required to confront and articulate these different conceptualizations of energy.

However, for students in multidisciplinary or interdisciplinary general science classes (such as middle-school science classes or those frequently taken by preservice elementary education majors at universities), the concept of energy appears in multiple contexts throughout the course. The term is likely used in many different ways, and different metaphors for energy are employed by the teacher and/or the textbook depending on the particular topic being studied. To complicate matters further, the words *energy* and *conservation* also have very different meanings in everyday discourse. Students in these courses face a difficult task; they must not only reconcile different conceptions of energy from a scientific perspective, but also distinguish between scientific and colloquial uses of the term.

*Purpose and Research Questions*

This study uses an analytical framework based on conceptual metaphor theory (described in detail below) to uncover the conceptual metaphors students used to describe the role of energy in various scientific contexts. Previous publications have explored how this framework could be used to understand how energy is conceptualized in pedagogical discourse (Lancor, 2014a) and by students in introductory physics, chemistry, and biology courses (Lancor, 2014b). This study differs from the previous empirical study in that the students were not explicitly asked to use analogies in their responses, and the students were enrolled in an undergraduate, interdisciplinary science course for non-science majors. This course focused on current issues in the news rather than a traditional sequence of topics covered in introductory science courses (e.g. mechanical systems or chemical reactions). There were two goals of this study: (1) to evaluate the methodological framework and determine whether or not it could be applied outside of a traditional disciplinary science course; and (2) to gain some insight into how these students, who are more or less representative of the general public, understand energy. Thus the research questions addressed in this study are:

- What conceptual metaphors are used by students in an interdisciplinary science course to explain the role of energy in various systems?

- How do they compare to metaphors of energy that have been previously identified in traditional scientific courses? Do students use the same conceptual metaphors for energy spontaneously as when they are explicitly asked to use analogies on energy?

## Literature Review

*Models, Metaphors, and Analogies*

Lemke (1997) argues that energy does not have one unambiguous definition that holds for all circumstances, but rather has a socially created meaning depending on the particular context of use. One way to make sense of these many interpretations of energy is through the lens of conceptual metaphor theory. Scholars of conceptual metaphor theory contend that many of our conceptual structures are built on metaphors, which help us to understand the world in terms of what is familiar (Lakoff & Johnson, 1980, 1999). Lakoff and Johnson (1999, p. 233) note that 'In the case of physics, there is certainly a mind-independent world. But in order to conceptualize and describe it, we must use embodied human concepts and human language.' When we encounter new ideas, we instinctively relate them back to what we already understand, which helps to make the new concepts intelligible.

As a part of our conceptual system, metaphors influence our perspective on the world. They do this by highlighting certain aspects of abstract concepts and obscuring others. When we conceptualize an experience or idea, we pick out the most important parts, find a way to categorize those parts in terms of what we already know about the world, and thus understand the experience. Within the field of science education, the theory of conceptual change recognizes metaphors (and analogies) as a key component of one's conceptual ecology (Posner, Strike, Hewson, & Gertzog, 1982). Metaphors and analogies 'help people explore their epistemological and ontological commitments' (Aubusson, Harrison, & Ritchie, 2006, p. 1). As researchers, we can work backwards; by analyzing the metaphors and analogies used by students to communicate their ideas, we gain insight into which ontological commitments they use to conceptualize energy in a given context.

Analogies and metaphors are often lumped together with models in the science education literature. Indeed, many times scientific models include analogies or metaphors (Aubusson et al., 2006) and mental models are often constructed through analogical reasoning (Collins & Gentner, 1987). Although they may not be aware of it, students harbor unique mental models that they use to explain the world around them. The difficult task for a researcher is to access these mental models. Hestenes (2006) describes three worlds: (1) the physical world where we interact with and observe phenomena, (2) the metal world where mental models are created to explain the phenomena, and (3) the conceptual world, the space in which mental models are communicated to others (often in the form of metaphors). The key to uncovering students' models is language—'Language does not refer directly to the world, but rather to mental models and components thereof!

Words serve to activate, elaborate or modify mental models, as in comprehension of a narrative' (Hestenes, 2006, p. 11). Thus understanding the language used to communicate the model allows us to understand the student's mental model and its relationship to the accepted scientific model.

Conceptual metaphor theory affords a systematic way to interpret this language. The idea of using metaphors as a tool to understand the world has much in common with scientific modeling. Generally speaking, scientific models predict and explain observed phenomena. They help us to make sense of new phenomena by making connections to, and expanding on, what we already know about how the world works. Similarly, metaphorical thinking is used to relate new ideas to prior experience. Additionally, scientific models simplify a system so that it may be described and quantified. Because of the simplifications required to create a scientific model, multiple scientific models are required to fully understand any given system. This is also true of the metaphors used to describe and explain a given system. Multiple coherent metaphors are necessary to gain a complete understanding of a system. This relationship between scientific models and metaphors had been noted by others (Duit, 1991; Hestenes, 2006). Viewing science as a set of coherent metaphors is not very different from thinking of science as a set of models; the way that we communicate scientific models is often metaphorical. Furthermore, multiple conceptual metaphors may be necessary to describe one scientific model, as is the case with energy, which will be explored in this paper.

Metaphorical construal of energy often involves what Lakoff and Johnson (1999) would call an 'Object Event Structure' metaphor (Amin, 2009). In particular, energy is an attribute of a system. A particular system may have a given amount of kinetic or potential energy. If we change the system somehow, the attributes of the system change. For example, if I drop a ball, the amount of kinetic energy increases while the amount of gravitational potential energy decreases. Typically, we say to students that the ball now *has* less gravitational potential energy at the bottom than it did at the top. In this way, we are conceptualizing energy as a possession of the ball. This is an example of what Lakoff and Johnson call the 'Changes are Movements of Possessions' metaphor, a subset of the Object Event Structure.

For the purposes of this analysis, we focus on substance metaphors as examples of the 'Changes are Movements of Possessions' mapping laid out by Lakoff and Johnson. Scientifically, we talk of energy being moved throughout a system or being transferred into or out of a system. However, it is difficult to speak intelligently about this movement of energy without connecting to an embodied experience of moving physical objects into or out of physical locations. The substance metaphors reflect the physical act of moving the energy substance from one system to another.

There is much discussion in the literature on how to define the terms metaphor and analogy. I use the term *analogy* to mean an explicit comparison of two ideas as expressed in written or verbal discourse. For example, a teacher may state 'The planetary model of the atom is like a solar system; the nucleus is like the sun and the electrons are like the planets orbiting the sun.'[1] *Metaphors* also compare two ideas, but do so implicitly. Additionally, I define *conceptual metaphor* to be the overarching ontological commitment that is supported by specific instances of metaphorical language and/or analogies. To summarize, the *conceptual metaphor* is how we interpret and apply scientific models, representing an underlying relationship between ideas; and *analogies* and *metaphors* are specific instances of discourse used to articulate those relationships (Table 1).

### Defining Characteristics of Energy

Energy is an abstract concept; it is not directly observable and is impossible to measure directly, which makes it difficult to define. Most scientists have a working definition of energy that is useful in their particular field, but is not broadly applicable. Undergraduate science majors have the opposite experience; many take multiple science courses concurrently and sift through various definitions of energy. Often students are expected to use the concept of energy in biology and chemistry before they have taken physics, and yet the definition given is based on physics principles (i.e. energy is the ability to do work). Research shows that students taking biology simultaneously with physics and/or chemistry are particularly confused by the concept of energy (Gayford, 1986). Additionally, attempts to illustrate the interdisciplinary nature of energy require simplifications that lead to nonsensical results (Zurcher, 2008). In an analysis of physics and chemistry texts, Taber (1989) found over 50 discrete manifestations of energy, some of which were synonymous, ambiguous, or simply incorrect. Many educators avoid this quagmire by simply never giving a definition of what energy is. When energy is defined by scientists, educators, or textbook writers, the definition typically falls into one of three categories: (1) energy defined through the concept of work; (2) energy as something that 'makes things go'; or (3) energy as a measure of change in a system.

Hand in hand with the debate about how to define energy, there is an extensive, ongoing debate about how best to teach the concept of energy (Jewett, 2008a, 2008b, 2008c, 2008d). Scholars agree that teaching the law of energy conservation alone is not enough to facilitate understanding of the complex concept of energy. Teaching conservation in tandem with transformation, transfer, and degradation leads to a more complete understanding of energy (Duit & Haeussler, 1994; Hecht, 2007; Nordine, Krajcik, & Fortus, 2011; Trumper, 1990). This list of characteristics was expanded to include energy source, as it has been identified as an important feature of energy in other studies (Lee & Liu, 2010). Taking the literature on energy instruction as a whole, five characteristics of energy have been identified and will be used in analyzing the students' written work:

- *Energy conservation*—In an isolated system,[2] energy can neither be created nor destroyed. This is one of several conservation laws used in physics.
- *Energy degradation*—The total amount of usable energy[3] in a system may decrease over time. This may take the form of energy dissipation (energy lost from an open system) or energy transformation within the system to a less useful form.

Table 1.   Definition and examples of models, metaphors, and analogy

| Term | Definition | Example 1 | Example 2 |
|---|---|---|---|
| System | The system includes all elements necessary to understand a given phenomenon. Open systems can exchange energy with their surroundings. Closed systems are (theoretically) isolated and do not exchange energy with the environment | Mechanical system (e.g. a ball rolling down a ramp) | Ecosystem |
| Scientific model | The scientific (or conceptual) model is used to explain a given phenomenon or gain understanding of some aspect of the system. Examples of explanatory scientific models include energy, momentum conservation, and natural selection | Energy: Gravitational potential energy is converted to kinetic energy as the ball rolls down the ramp | Energy: The energy inputs and outputs of a system can be tracked and used to determine rates of production and consumption |
| Metaphor | Metaphorical language is used to explain the scientific model in more concrete terms, and implies a relationship between the target concept and some more familiar concept | Energy is the currency of the system. The ball has the same *total amount* of energy at the bottom of the ramp as it had at the top | Energy pours into an ecosystem as solar radiation and drains away as respiratory heat loss (Campbell & Reece 2002, p. 1206) |
| Analogy | An analogy is used to explain the scientific model in more concrete terms, and explicitly states a functional or structural relationship between the target concept and the analog | Energy is like money. The ball has a set amount of energy at the top; this is the potential energy. Imagine you have $10 in your pocket. If you go to the bank and deposit the 10-dollar bill, you still have $10, but it is in a different form—now it is in the bank account instead of your pocket. This is like the potential energy being converted to kinetic energy | Energy flows through an ecosystem like water flows through an irrigation pipe. Some makes water through to the next field, and some leaks out of the system |

*(Continued)*

Table 1.   Continued

| Term | Definition | Example 1 | Example 2 |
|------|-----------|-----------|-----------|
| Conceptual metaphor | The conceptual metaphor represents the overarching relationships between components in the target concept and the source domain. The conceptual metaphors are based on specific instances of metaphorical language or explicit analogies identified in discourse | Energy is a substance that can be accounted for. This conceptual metaphor highlights the principle of energy conservation, and gives us a way to track changes in energy in a system | Energy is a substance that can flow. This conceptual metaphor emphasizes the idea of energy transfer through a system |

- *Energy transformation*—Energy can be transformed from one form to another. For example, as a ball drops gravitational potential energy is transformed into kinetic energy.
- *Energy transfer*—Energy can be transferred between components in a system; in a collision, one billiard ball transfers its kinetic energy to another.
- *Energy source*—Energy can be added to a system. For example, in an ecosystem, an input of energy from the sun is needed to balance the loss of thermal energy from the ecosystem to the environment.

## Metaphors for Energy

The framework for evaluating student ideas about energy was developed based on a survey of written materials from biology, chemistry, and physics, including text-books[4] and the science education literature. Specific examples of metaphorical language and explicit analogies were identified following the method presented by Lakoff and Johnson (1980, 1999). These instances were then grouped into themes, representing variations on the Object Event Structure metaphor laid out by Lakoff and Johnson. Themes were identified that represented similar ways of understanding the role energy plays in a system (e.g. it can be stored, it can change forms, etc.). Generally the metaphors fall into the categories of either 'Attributes are Possessions' (i.e. the ball has kinetic energy; there is no change in the system) or 'Changes are Movements of Possessions' (i.e. Ball A transferred some of its energy to Ball B during the collision; there is a change in amount of energy possessed by each ball).

Note that no metaphor is exclusive to any one discipline. Each discipline may use one metaphor preferentially, but the other metaphors certainly make appearances. The goal here is not only to highlight the differences among disciplines, but also to recognize that common conceptual metaphors are being used across disciplines. The language may seem different on the surface, but the underlying relationships are similar. The conceptual metaphors are described in brief below, as well as some discussion of how they map onto the characteristics of energy defined above (Table 2). A more detailed account of the development of the framework can be found elsewhere (Lancor, 2014a).

*Energy as a Substance that Can be Accounted for*

References to energy as a substance that can reside in various 'accounts' or 'containers' are common in both physics and chemistry texts, such as toy blocks (Feynman, Leighton, & Sands, 2006) or money (Chang, 1998; Knight, 2007). The amount of energy in each 'account' changes as a result of some interaction with another system. These examples illustrate the conceptual metaphor that *energy is a substance that can be accounted for* within a given physical system. This metaphor is reinforced

Table 2. Conceptual metaphors identified in biology, chemistry, and physics discourse. The metaphor represents the overarching framework, supported by explicit analogies that highlight or obscure characteristics of energy

| Conceptual metaphor | Examples of analogies from scientific contexts | Characteristics of energy | |
|---|---|---|---|
| | | Highlights | Obscures |
| Energy as a substance that can be accounted for | Energy (or enthalpy) is like money | Conservation | Transformation |
| | Energy is like a child's blocks | | Source |
| Energy as a substance that can change forms | Solar energy converted into chemical energy through photosynthesis | Transformation | Transfer |
| | Chemical energy converted into thermal energy in an exothermic reaction | Conservation | |
| Energy as a substance that can flow | Energy flows through an ecosystem | Transfer | Transformation |
| | Heat flows from hot to cold | Source | |
| | Electricity flows through a circuit | | |
| Energy as a substance that can be carried | Organisms transport energy through an ecosystem | Transfer | Transformation |
| | Photons carry electromagnetic energy | | |
| Energy as a substance that can be lost | Trophic pyramid | Degradation | Conservation |
| | Energy is lost in an exothermic reaction | Source | |
| Energy as a substance that can be added, produced, or stored | Energy is stored in chemical bonds (e.g. ATP) and can be released | Source | Conservation |
| | Energy is stored in a capacitor | Transfer | Degradation |
| | Energy is added to initiate a chemical reaction | | |

through graphical representations like bar charts (Scherr et al., 2012). The accounting system metaphor gives scientists a tool to apply energy conservation quantitatively, to track energy changes and interactions between systems. This accounting system is useful because it portrays energy as a substance that can be tracked. This conceptual metaphor emphasizes the conservation and transfer aspects of energy, but obscures the idea of energy transformation. The 'energy' in these examples is generally of the same form—it is a block or money—it never changes to another form of energy (e.g. a block does not change into a ball).

## Energy as a Substance that Can Change Forms

The 'forms of energy' language is ubiquitous in science texts. It is generally accepted that these forms of energy fall into two broad classes: kinetic energy, which involves motion, and potential energy, which is stored in fields. Many scholars do not see a problem with the 'forms of energy' language, arguing that if used correctly this metaphor can represent a scientifically accurate understanding of energy (Kaper & Goedhart, 2010; Nordine et al., 2011; Trumper 1990). Obviously, this metaphor highlights the transformation of energy, particularly when used in tandem with the conservation principle. For example, 'If one form of energy in an isolated system decreases, then another form of energy in the system must increase' (Serway, Faughn, & Vuille, 2006, p. 118). In this way, the 'forms of energy' metaphor is a heuristic that helps to explain how energy is conserved in various situations. The 'forms of energy' metaphor can be used in conjunction with the accounting system metaphor; the forms of energy could be construed to be the various 'accounts' discussed above. According to the principle of energy conservation, we can never destroy or lose energy in an isolated system; if energy appears to be missing, scientists will search for another 'form of energy' that may account for the missing energy (as in the current search for dark energy). On the other hand, this metaphor obscures the transfer of energy; it provides no explanation for how energy can be passed from one object to another without changing forms.

## Energy as a Substance that Can Flow

The metaphorical phrase 'energy flow' makes one imagine a pipe with water flowing through it. Energy flow language is used repeatedly in biology, chemistry, and physics textbooks. For example: 'Energy flows through ecosystems, while matter cycles within them' (Campbell & Reece, 2002 p. 1198); 'we often speak of "*heat flow*" from a hot object to a cold one' (Chang 1998, p. 205); and water flow analogies in the context of electrical circuits (Harrison & Coll, 2008). This language highlights energy transfer in a system. The 'water' (energy) substance stays the same in this metaphor, in contrast to the 'forms of energy' metaphor described above where energy takes on a different form as a result of an interaction in the system. Thus this metaphor highlights the transfer of energy while downplaying energy transformation. And if energy flows into a system, it has to come from somewhere outside the system, an external source of energy. The flow metaphor is a convenient way to discuss a continuous, uniform, energy transfer through a system.

### Energy as a Substance that Can be Carried

Energy can also be conceptualized as a substance that can be contained and carried. For example, an electron 'carries' energy through an electrical circuit; organisms 'transport' energy through ecosystems. Both the electrons and the organisms could be considered to be energy carriers. Falk, Herrmann, and Schmid (1983) advocated for language of energy carriers, arguing that it is more scientifically accurate to view an energy transformation as energy being transferred from one carrier to another. For example, rather than saying the chemical energy in a battery is converted to electrical energy in a circuit, we would say a battery carries a given amount of energy, and then passes that energy along to an electron, which carries it through the circuit. Rather than thinking of the energy as changing form, the energy has a different carrier.

### Energy as a Substance that Can be Lost from a System

The metaphor of energy as a substance that can be lost from a system is prevalent in biology textbooks, particularly in the discussion of ecosystems. In this context, the systems of interest are primarily open systems in which thermal energy is freely transferred to the surrounding environment. For example, 'on average, these primary consumers harvest 31 kcal/m$^2$ of energy each year. Of that total, 17.7 percent is unused and excreted and 80.7 percent is lost to respiration and other maintenance processes' (Freeman, 2007, p. 1230). This is more aligned with how students hear about energy in the media (e.g. turn off the lights because we are running out of energy) than it does the scientific notion of energy conservation. The 'energy loss' metaphor does a fantastic job highlighting energy degradation, but obscures energy conservation. For this reason, scholars have argued that degradation needs to be taught in parallel with energy conservation (Duit & Haeussler, 1994; Pinto et al., 2005).

### Energy is a Substance that Can be Stored, Added, or Produced

In any chemical reaction, an input of energy is necessary to break bonds. Whether energy is absorbed or released overall depends on the particular reaction and the differences in binding energy between the ingredients and products. However, students often hold the misconception that energy is released when bonds break (Boo, 1998). Language indicating energy as an ingredient or a product is common among students (Trumper, 1990; Watts, 1983). Unfortunately, this can lead to confusion between matter and energy in chemical reactions (Anderson 1990) and ecosystems (Barak, Sheva, Gorodetsky, & Gurion, 1999; Leach, Driver, Scott, & Wood-Robinson, 1996; Lin & Hu, 2003). The language used to describe the role of energy in chemical reactions reinforces this idea and reflects a conceptual metaphor that energy is either an ingredient or a product of a reaction. However, this does provide the means to discuss the concepts of energy transfer and energy source in a meaningful way with students, provided it is emphasized as a heuristic metaphor for understanding the role of energy in facilitating chemical reactions.

A related idea is that energy can be stored. Both chemistry and biology texts describe how energy can be stored in bonds, even though this idea is commonly considered a misconception (Gayford, 1986; Novik, 1976). In many cases, a chemical bond is equated to a loaded spring (Campbell & Reece, 2002) or a battery (Harrison & Coll, 2008). The energy storage language is useful in discussions of potential energy in general (Swackhamer, 2005). Physics classes abound with language of energy stored in batteries, springs, even a block at the top of a ramp that 'stores' gravitational potential energy. The energy storage language is common among younger students as well (Watts, 1983).

*Are Substance Metaphors Valid?*

One finding from this analysis is that the vast majority of discourse about energy implies that it is a substance. Although widely accepted that energy is not actually a substance, it is virtually impossible to discuss energy without referring to it as a tangible quantity. These metaphors are not only common, but also provide a fruitful framework for helping students conceptualize the abstract notion of energy. Any ontological metaphor either highlights or obscures the various aspects of a given concept. In this case, the fact that energy is not a substance is obscured so that the other characteristics may be made clear. The downside is that this language implies that energy is a physical substance. Even so, many educators recognize that substance metaphors are not harmful to students' understanding of energy (American Association for the Advancement of Science, 2008; Duit 1987; Falk, Herrmann, & Schmid, 1983).

Although many of these conceptual metaphors are commonly cited as alternative conceptions (Watts, 1983) or fallacies (Sefton, 2004) students have about energy, it is probably more accurate to say that students holding these ideas have an incomplete understanding of energy. A complete definition of energy would recognize that energy is a conglomerate of the ideas listed above. Energy can flow through ecosystems; it can be the product of a reaction, and so on. None of these is entirely correct on its own, but each highlights different aspects of the broad concept of energy as they are used in a particular context. Taken as a whole, they form a set of *coherent* conceptual metaphors for energy. The value in each conceptual metaphor is that it helps to explain the role of energy in its application to a particular context; energy cannot be defined out of context or outside of a system.

It is important to recognize that there are limitations to using substance metaphors for energy (Amin, 2009; Scherr et al., 2012). The primary one being that it is difficult to conceptualize negative energy as a substance, as is the case for electrical or chemical potential energy in bound systems. For this example, some have pointed out that in this context it is more fruitful to use the Location Event Structure, as defined by

Lakoff and Johnson (1999), in which energy is described as a physical location (e.g. *up* or *down*) rather than a tangible substance (Dreyfus et al., 2014).

To summarize the literature, there is no consensus as to which definition (or metaphor) of energy is best. One consequence of this is that students receive mixed messages during instruction, and the definitions of energy are often at odds with each other. By documenting the different conceptual metaphors students use to describe energy, we can begin to understand the effect that these different conceptualizations have on student learning.

## Methods

### Participants and Setting

The participants in this study were enrolled in a university two-semester interdisciplinary general science course for non-science majors. Content was covered in an integrated manner using a Science, Technology & Society approach. A student who completed the course was expected to be able to read, understand, and intelligently discuss science-related stories in the media. Topical units included issues such as human energy use, transportation, radiation, natural disasters, and space exploration (in the first semester). The fact that energy appeared throughout the course gave students multiple points of entry for understanding this complex concept, and provided a unique environment for research because the students were exposed to the various scientific meanings of the term within one course.

The course was taught at a small liberal arts college in the Midwestern United States, and enrolls either 20 or 40 students depending on whether one or two sections are offered. The course met for two 75-minute lectures per week and one 3-hour laboratory session. Data presented in this paper were collected over two years in the first semester of the course sequence. All students taking the course were invited to participate in the study. A total of 49 students participated in the study over the two years. Students came from a range of majors (mostly business, early childhood education, or the humanities), and were generally juniors and seniors. These students typically had taken only two years of high school science, and no other college science course. Thus these students can give us some idea of how the general public understands energy.

### Data

The data reported in this paper were drawn from an essay question on the final exam. This was a take-home exam; students were permitted to use notes, textbooks, and other resources. The question asked:

Energy and energy conservation (in the scientific sense) are key to understanding most topics in science. Look at the list of topics below and explain how energy and/or energy conservation are involved and why energy is important in understanding this topic.

(a) Radiation/radioactivity (e.g. taking an x-ray),
(b) Generating electricity using fossil fuels,

(c)  Transportation (e.g. using gasoline to fuel a car),

(d)  Earthquakes and tsunamis, and

(e)  Creation of the universe (e.g. big bang).


*Analysis*

The data analysis draws on the methodology used by Lakoff and Johnson (1980, 1999). In their work, Lakoff and Johnson identified metaphorical phrases in language and grouped them together by theme. For example, the phrases 'Look how far we've come', 'I don't think this relationship is going anywhere', and 'We're at a crossroads.' reflect the conceptual metaphor LOVE IS A JOURNEY (Lakoff & Johnson, 1980, p. 44).

In the previous study (Lancor, 2014b), the students' analogies were grouped together based on a method of constant comparison, eventually converging on a list of six conceptual metaphors (described above). In this study, explicit analogies and instances of metaphorical language were identified and classified according to the following criteria:

- *Energy as a substance that can be accounted for.* Evidence included mention that the amount of energy in the system (changing or staying the same) could be counted. For example, 'X energy is here, Y energy is there, but the total amount of energy stays the same.'
- *Energy as a substance that can change forms.* Language indicating energy can change forms included 'X changes into Y', 'X is converted into Y', and 'X is transformed into Y'.
- *Energy as a substance that can flow.* Any language indicated a fluid movement of energy was coded as a flow metaphor. This included energy 'flowing' out of or through a system.
- *Energy as a substance that can be carried.* Anytime a student wrote that energy was moved from one location to another *by an object* was coded as an energy carrier metaphor. Verbs like carried, held, or transported were common. Also, if a student wrote an object 'has energy in it', it would be coded as an energy carrier metaphor.
- *Energy as a substance that can be lost.* Evidence of this metaphor included language such as 'energy is lost from the system' and 'energy is no longer useful or usable'.
- *Energy as a substance that can be added, produced, or stored.* Evidence of this metaphor included students writing that 'energy is needed for X to happen'. Other verbs included energy being created, produced, stored, added, or released.

In this study, many students used multiple metaphors to explain a given scientific topic. In these cases, the discourse was coded for both metaphors (sometimes even three metaphors).

For example:

> Also, the plate boundaries slide against each other, which will cause earthquakes, so the heat and gas energy will release in the environment. The heat and gas energy are not lost because it goes into the atmosphere.

144

The statement that energy will be released into the environment indicates an energy storage metaphor (if it was released, it had to be stored somewhere). The student also states that the energy is 'not lost', which is an indication of the accounting system metaphor and reflects an understanding of energy conservation.

A second round of coding determined which of the characteristics of energy were present in the students' responses. The criteria for each characteristic were:

- Energy conservation. Responses were considered to have evidence of energy con- servation if they discussed a fixed amount of energy, or recognized that energy is never lost, destroyed, or created.
- Energy degradation. Evidence of energy degradation included recognition that energy can be lost from a system or that the total amount of (useable) energy in a system decreases.
- Energy transformations. Evidence of transformation was primarily that the analog to energy had the ability to change forms. Some students used the word 'transform- ation' but their analogy did not actually indicate that the substance changed forms in the analogy (e.g. water (energy) being poured from one bucket to another). These were not coded as transformation, but as energy transfer.
- Energy transfer. Evidence of energy transfer included the substance (energy) moving from place to place or being transported by an agent.
- Energy source. A clearly identifiable source of energy was included in the response (e.g. 'the ocean represents the sun, a source of energy').

The goal was to determine whether or not students used these characteristics in explaining the concept of energy. Note that student responses did contain many of the misconceptions cited in the literature (e.g. energy is causal, an anthropo- centric view of energy). The focus of this analysis is on how students use meta- phorical language to explain, and therefore conceptualize, energy. This is intentionally not an evaluation of the scientific accuracy of their claims. The goal of the study is to learn more about student cognition, not to evaluate mastery of the science content. Some of the examples cited below do contain scientific errors, but I have generally refrained from discussing these misconcep- tions in the analysis.

## Results

The results presented below are grouped by conceptual metaphor to highlight the similarities and differences in the way that the metaphors were used in the various scientific contexts. Virtually all of the metaphors used by students could be placed into the previously identified categories. Table 3 compares this data to the results from the previous study which identified metaphors in traditional science courses. The overall results of coding for the characteristics of energy are shown after the meta- phors are discussed.

Table 3. Prominent metaphors students commonly used to describe energy in this study and from the previous study on topics from traditional science courses (Lancor, 2014b)

| | | | Metaphors for energy (% of student responses) | | | |
|---|---|---|---|---|---|---|
| Topics from interdisciplinary science course[a] | Energy as a substance that can be carried | Energy as a substance that can flow | Energy as an ingredient, product, or substance that can be stored | Energy as a substance that can change forms | Energy as a substance that can be lost | Energy as a substance that can be accounted for |
| Radioactivity | 25 | 25 | 25 | 13 | 6 | 0 |
| generating electricity | 6 | 0 | 54 | 27 | 12 | 0 |
| Transportation | 3 | 3 | 25 | 47 | 22 | 0 |
| big bang | 23 | 15 | 31 | 8 | 0 | 23 |
| Earthquakes and tsunamis | 13 | 24 | 45 | 16 | 0 | 3 |
| Topics from traditional science courses[b] | Energy as a substance that can be carried | Energy as a substance that can flow | Energy as an ingredient, product, or substance that can be stored | Energy as a substance that can change forms | Energy as a substance that can be lost | Energy as a substance that can be accounted for |
| Mechanical systems | 11 | 10 | 9 | 22 | 0 | 48 |
| Circuits | 19 | 70 | 4 | 7 | 0 | 0 |
| Ecosystems | 15 | 38 | 12 | 5 | 25 | 5 |
| Chemical reactions | 6 | 4 | 58 | 8 | 0 | 14 |

[a]Totals may add up to more than 100% because some responses were coded for multiple metaphors.

[b]In the previous study, the analysis included a code for 'process metaphors' which was not included in the current study. For this reason, not all totals add to 100%.

## Energy as a Substance that Can be Accounted for

The accounting system metaphor was practically nonexistent in the student responses, which indicates that students may not see this metaphor as useful in describing energy in these scenarios. It is interesting that this metaphor appears so rarely because it was used extensively in the discipline-based classes, particularly in the physical sciences (Table 3).

## Energy as a Substance that Can Change Forms

Energy transformation is considered by some scholars to be a hallmark of understanding the concept of energy (Nordine et al., 2011; Trumper, 1990). In this study, many students identified forms of energy (e.g. kinetic energy and thermal energy), but did not show evidence of understanding that energy can be transformed from one form to another. Language describing various forms of energy and language describing energy transformation both fall under the umbrella of this metaphor, but the latter represents a more sophisticated understanding of the concept of energy. Examples of both forms of energy and transformation language are given in Table 4 to illustrate the difference in complexity of these responses.

The forms of energy language is common in the context of generating electricity and transportation because it emphasizes the many energy transformations that take place as energy propagates through these systems. The energy transformation metaphor is a useful framework for understanding complex systems with many interacting parts. On the other hand, the transformation metaphor appeared rarely in the context of either radiation or the big bang (only 2 and 1 responses, respectively). This is likely not a useful metaphor in those two contexts because it does not emphasize the characteristics of energy that are most important to explain these phenomena (i.e. energy transfer rather than energy transformation).

## Energy as a Substance that Can Flow

The metaphor of energy as a substance that can flow (like water) occurred in the context of earthquakes and radiation, but was not used frequently compared to its prominence in the previous study (Table 3). For example:

> When these earthquakes occur in the ocean, the ground movement causes a wave that the energy flows through.

In this example, the energy flows through the wave; in other examples, energy is carried by a wave, making the wave an energy carrier. The metaphor of energy waves as a mode of energy transfer is an interesting case, and could be considered a subset of either the energy flow metaphor or the energy carrier metaphor, or perhaps a conceptual metaphor in its own right. Further investigation needs to be done to determine the extent to which this metaphor is useful in a range of contexts.

Table 4.  Examples of student responses with evidence of the *Energy as a substance that can change form* metaphor

| Context | Excerpts of student responses (emphasis added) | |
| --- | --- | --- |
| | Forms of energy | Energy transformation |
| Generating electricity | First off, electricity is a <u>form of energy</u> and can be generated by many sources such as hydro-electric energy | At power plants, coal is burned and its potential chemical energy heats up water in a boiler. When the water boils, it releases thermal energy in the form of steam. Then the steam powers a turbine engine by <u>transforming</u> the heat into kinetic energy that spins the turbine engine. After that, the turbine engine uses the kinetic energy to power a generator. The generator finally takes the kinetic energy and <u>transforms</u> it into electrical energy. Throughout this process, it can be understood that energy is primarily lost in the form of heat |
| Transportation | There are many <u>types of energy</u> involved in transportation. There is the most obvious which is seen in the movement of the car or Kinetic energy. The less obvious are those within the vehicle itself. <u>There is electrical energy, light energy, thermal energy, chemical energy, gravitational energy, potential energy and friction.</u> Chemical energy is in the burning of oil and the gas/fuel, and the battery. All of the forms of energy are needed to make a car run | The gasoline serves as the potential chemical energy that eventually <u>turns into</u> kinetic energy to place a vehicle in motion. During this energy <u>transformation</u>, the heat from the engine breaks down the chemical bonds in the gasoline. Then when these bonds break, their chemical energy is released and places the gears inside the vehicle into motion when a person pushes their foot down on the gas pedal to drive. However, not all of the gasoline's energy goes into powering the vehicle. A lot of it becomes lost in the form of heat and sound when the vehicle's engine and gears are working |
| Earthquakes and tsunamis | Energy involved with earthquakes and tsunamis include <u>kinetic energy, friction, and geothermal energy</u> | Earthquakes are caused by two tectonic plates rubbing against each other and creating potential which is waiting to be released. Once this energy is released it is <u>transformed</u> into kinetic energy |

## Energy as a Substance that Can be Carried

Under this metaphor, an energy transformation is explained as energy being transferred to a different carrier; the new carrier becomes the vehicle for what would be a different form of energy in the previous metaphor. This language is not frequently used in standard scientific discourse, but the students found this to be a useful metaphor. For example (emphasis added):

> Radiation itself is a type of energy that is <u>packed into small units</u> called Alpha and Beta particles.

> X-rays are basically the same thing as visible light rays as they are both forms of electromagnetic energy <u>carried</u> by particles called photons.

> The [tectonic] plates <u>carry</u> potential energy and when they shift, the energy they carry gets transferred into the ground creating a shaking from the kinetic energy which can and many times results in an earthquake.

Note in the earthquake example, the energy is transferred between multiple carriers, which would be a transformation under the previous metaphor.

*Energy as a Substance that Can be Lost*

The metaphor that energy is a substance that can be used up or lost was common in student responses, particularly in the two scenarios involving fossil fuels: generating electricity and transportation. Many students recognized fossil fuels (and also energy) as finite natural resources that must be conserved. Many students also recognized that the energy is not really lost, but is degraded into a less usable form or transferred to the environment. For example:

> The problem is that the input of energy from coal is almost equal to that of the energy which is transferred elsewhere or 'lost' during the conversion of this stored energy into electric. If the electricity production seems bad, then the energy usage of a car is worse. Energy that is lost to the system is no longer useful, resulting in degradation of the energy.

However, not all students recognized that energy was lost *from the system*. In these cases, it is not obvious that students recognize that the energy goes somewhere else; it could just be disappearing. For example, this student implies that the amount of useful energy (in the gasoline) in the system is decreasing, but does not recognize where that energy goes:

> … With normal gasoline this energy is lost, but with a biodiesels the energy from carbon dioxide goes back into the planets grown for the fuel, that why it's important to understand how this energy works so we can better our environment.

This makes it difficult to tell if the student has an understanding of the relationship between energy conservation and energy degradation. The distinction between these two variations on the energy loss metaphor is important because it draws a line between a student with a misconception (i.e. energy can be destroyed) and a student with a more complete understanding of how useful energy can dissipate from a system.

*Energy as a Substance that Can be Added, Produced, or Stored*

This is another metaphor that appeared frequently in each of the topics. It is interesting to see that this metaphor plays out in very different ways in this diverse group of scenarios (Table 5). Whether the energy is an ingredient or a product often depends on how the system is defined. Additionally, language of energy storage is often intertwined with language indicating energy is an ingredient or a product. For example, energy is *produced* from the combustion of gasoline (the system is the fuel), but is

Table 5.   Example of student responses with evidence of the *Energy as a substance that can be added, produced, or stored* metaphor

| Context | Example of Student Responses (emphasis added) | | |
| --- | --- | --- | --- |
| | Ingredient | Product | Storage |
| Radiation | When taking an x-ray, high energy photons are <u>needed</u> to produce the x-ray. Inside of the x-ray vacuum, electrons are constantly jumping between energy levels, releasing an x-ray photon each time | These photons are <u>produced</u> by the movement of electrons in atoms. Atoms emit light by colliding with a moving particle which in turn causes an electron to climb to a higher energy level, and the fall back to its original level. This causes the extra <u>energy to release</u> in the form of a light photon | Your bones <u>absorb</u> this energy much better than the tissue in our bodies |
| Transportation | When gasoline is used to fuel a car it goes through a combustion engine and <u>provides</u> energy for the motor to move the wheels of the vehicle | Gasoline, and the energy it <u>produces,</u> is very important because it provides us with a way of transportation | Some motor vehicles are powered by gasoline, which <u>holds</u> potential energy |
| | | When the car is moving forward, there is kinetic energy involved. Also, it <u>produces</u> thermal energy when it has friction with the tires and the road | Chemical energy is <u>stored</u> in gasoline which you use to power your vehicle |
| Earthquakes and tsunamis | That earthquake released a lot of (energy) that started to <u>generate</u> waves out at sea | As the plates slide along one another and collide, <u>energy is created</u> between the frictions and grinding of the two plates | Earthquakes are caused by the rapid <u>release of stored</u> energy (potential energy), turning into movement (kinetic energy) ... This sudden release of energy then causes the ground to shake |
| | Geothermic energy is what creates earthquakes and tsunamis | | |

an *ingredient* needed for the car to run (where the system is the car). The energy is also *stored* in the gasoline in the tank, waiting to be combusted. Students write about energy production saying that energy is made, created, emitted, released, or produced. There are subtle differences; if energy was released that implies that it must have been stored in some way, if energy is created from some other type of energy that implies transformation. In general, this metaphor illustrates that energy can be transferred into or out of a system.

The fact that students used the energy storage metaphor repeatedly is interesting because it was not prevalent in the disciplinary contexts.[5] The idea of energy storage is commonly considered a misconception, and as a consequence the metaphor is not often used in the discourse of traditional disciplinary science courses. (And often it is explicitly addressed as an incorrect way of conceptualizing the energy in chemical bonds.) The fact that energy storage is commonly invoked here may reflect a lack of disciplinary expertise, and indicate that students are more likely to use intuitive metaphors for the more familiar, real-life scenarios. As was common in the course, students were constantly switching between the scientific and colloquial discourses and may not have made clear conceptual distinctions between them.

*Characteristics of Energy*

The purpose of identifying conceptual metaphors is to figure out which characteristics of energy the students understand. These results are shown in Figure 1. It is interesting to note that the profiles of each topic are quite different, but that energy transformation or energy transfer was the most evident in students' responses. Students may not be using energy conservation explicitly as a scientific model, but they are using the fact that energy changes (either form or location) in a system to explain these phenomena.

A final observation on the students' responses in this study is that there was generally less evidence for the characteristics of energy than in the responses from the students in traditional disciplinary science courses (Figure 1). This may indicate that a disciplinary structure helps students to have a more multifaceted understanding of the energy concept. The concept of energy is so abstract that it is difficult to conceptualize outside of a well-defined set of disciplinary norms. In an interdisciplinary science class, teachers need to be aware that the students' conceptions of energy are more fragile than those of students in a disciplinary course. They may have more difficulty piecing together the various characteristics of energy because they see the concept of energy used in so many different ways.

## Discussion

This study demonstrates that students use metaphorical language spontaneously (without being prompted) to describe energy, and furthermore the metaphors in the students' responses are drawn from the same set of conceptual metaphors identified in traditional, disciplinary science courses. This shows that the methodological framework developed for the original study is useful for analyzing discourse in this

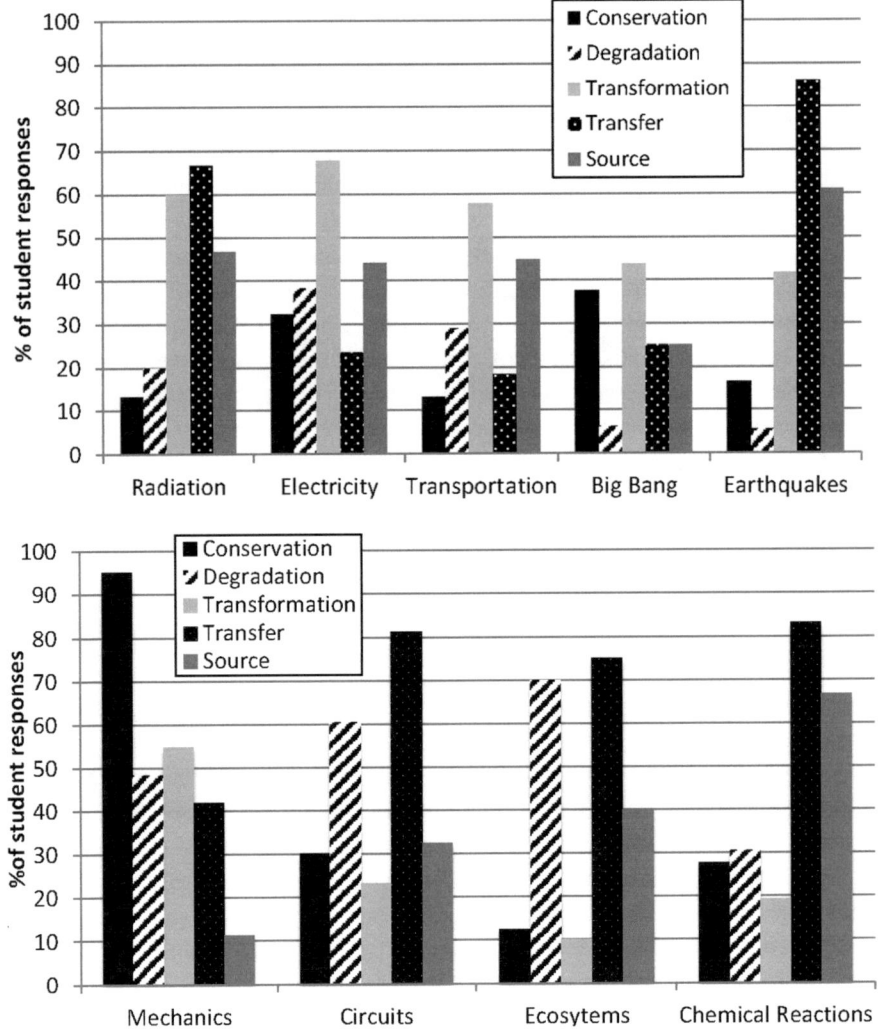

Figure 1.   Percentage of student responses that indicated an understanding of each characteristic of energy in various scientific contexts from this study (a) and the previous study (b)

context. As a whole, these students were far from being steeped in the discourse of a particular discipline, but the language they used had many commonalities with disciplinary discourse about energy. This is significant because these students had not taken any other college-level science courses, and many only had two years of high school science (the state graduation requirement). The implication is that we, as teachers, can use this framework to identify conceptual metaphors for energy, and use these metaphors to assess how students understand the various characteristics of energy in a wide range of science courses.

Comparing these results to the previous study (Table 3), there are some interesting differences in which metaphors were used preferentially. The metaphors of energy as an accounting system and energy flow through a system are prevalent in traditional science contexts and uncommon here. Energy loss and energy as an ingredient/ product/storage metaphors were much more common in the topics studied here, possibly because they are more common in everyday discourse about energy. Another observation to make about the data in Table 3 is that the students used a wider variety of metaphors to explain the topics in this study, as opposed to the findings from the earlier study that each scenario had one or two metaphors that were used predominantly.

Overall, applying the framework in this study was not as straightforward as when students were explicitly asked to write analogies. Student responses were often coded for multiple metaphors due to the complexity of their explanations. For example:

> Energy is also present when generating electricity. Much of the nation's electrical energy comes from coal. ... In order to get electrical energy from coal it must go through a process at a power plant to get the potential energy out.

This first part of the student's response indicates that energy is stored in the coal, and would be a product of the combustion reaction. The student is using the ingredient/ product/storage metaphor. She goes on to say:

> The coal is combusted using a large amount of heat to form steam. A steam turbine then converts energy from the moving steam into mechanical energy. The electrons are captured instead of being used immediately so they can be sent to people around the nation to use for electricity. Electricity is a form of energy. This is why there is an importance to understanding energy when talking about and dealing with electricity. You can't have electricity without energy.

The second half of her explanation uses the energy transformation metaphor; energy is converted from thermal to kinetic energy. She also implies that the electrons are carries that will transport energy to people's homes. She has used three metaphors simultaneously to highlight different aspects of the process of generating electricity.

In the following example, the student used an explicit analogy to explain the role of energy in generating electricity:

> Energy is something like a soul; it cannot be destroyed and is always around. However, unlike our souls, energy can be transferred to different activities. Think in a way that makes [sense] such as when you eat your [sic] absorbing or transferring that energy that was contained in that apple or orange and is now in your body.

This particular student starts by comparing energy to a soul, but recognizes the limitation of this analogy—the soul cannot be transferred. He then goes on to compare energy to food, using the metaphor of an energy carrier. The fruit is the carrier, transporting the energy to the person. However, he also finds a limitation because he cannot explain what it is in the fruit that is the energy (what is the substance (energy) being carried by the food?). He goes on to write:

> When we conserve energy it is like putting water into a jug with a tap, like in your kitchen. We have a way of getting that energy easily but at times we need to refill it because we notice that the water is almost gone. This water jug is like fossil fuels, except when

> we're running low we cannot just go and refill the water jug from our close sink. We need to walk a tremendous distance to a water spring in the jungle. This trip would take us a year to make just to fill our jug back up, but in our world fossil fuels are the water and the spring we get it from is formed over millions of years, so we can't just make easy trips.

In this analogy, energy is a substance (water) that is contained by the water jug (the fossil fuels). In this response, we can see that his multiple analogies are creating a more complex picture of the energy concept than could be achieved by using only one metaphor.

As the examples above illustrate, students often used two or more metaphors simultaneously. It may be tempting to dismiss these mixed metaphors as incomplete or incorrect understandings. However, under conceptual metaphor theory, we expect complex ideas to be represented by a set of coherent metaphors, which are necessary to highlight the different characteristics of energy. Interestingly, multiple metaphors were used more extensively here than in the traditional science courses. This could be due to the fact that these were not simple systems, unlike the ones examined previously (e.g. mechanical systems or circuits) and multiple representations of energy are required to adequately describe the phenomena.

Asking students directly to define energy tells us little about how they actually understand the concept, and does not help us to gain insight into student ideas. This is why we do the metaphor analysis—when asked for a formal definition of energy, students can state the Law of Energy Conservation, but we do not actually have evidence that they are understanding this concept or find it fruitful to explain the role of energy in a system. The analysis of the students' writing using conceptual metaphor theory can help a teacher assess how well students actually understand the energy concept, as opposed to evaluating how well they can parrot back formal definitions.

*Implications for Instruction*

Qualitative discourse analysis has typically been relegated to the world of science education research, and not practiced systematically by classroom teachers. One goal of this project was to develop a framework that teachers could use to help them interpret student ideas about energy. As such, the substance metaphors for energy described above are designed to be accessible to classroom teachers as well as educational researchers. This qualitative metaphor analysis would not be a good summative assessment tool because there is no one right metaphor for energy that is scientifically accepted in any context. Rather, we can use this framework as a formative assessment tool to help teachers interpret classroom discourse (spoken and written).

One of the purposes of formative assessment is to monitor the progress of students' conceptual development. Research shows that formative assessment helps to improve learning, but only if teachers use the data gathered by formative assessments to influence their teaching (Black & Wiliam, 1998). Black and Wiliam (1998) note that discourse and questioning can be powerful tools for formative assessment, but the students' responses can be difficult to interpret, and therefore difficult to use in making instructional decisions. Additionally, Bell and Cowie (2001) note that one of the characteristics of good formative assessment is that it be an integral part of teaching and learning. There is a need for more authentic assessments, assessments that are integrated into the curriculum (Tamir, 1998).

As a formative assessment tool, the goal of the metaphor analysis described above would be to help the teacher identify which characteristics of energy are articulated in the students' metaphors. This allows teachers to better build on the students' existing ideas. Energy is a complex concept that can only be successfully applied if a learner has mastered all of it various characteristics (e.g. conservation and transformation). Information about which characteristics students understand, and which they do not, is valuable information to have as teachers plan future lessons. Using this method of discourse, analysis yields a more nuanced picture of student understanding than can be gained from traditional assessment questions such as 'What is the definition of energy?'

*Limitations of the Study and Future Work*

One limitation of this study was the exclusive focus on substance metaphors. It is quite possible that students used other metaphors for energy, but analyzing the responses for other metaphors was beyond the scope of this study. A future research project could re-analyze the data to look for evidence of the 'Location Event Structure' metaphors. Additionally, future studies could take a longitudinal approach, and examine how students' ideas about energy develop over time. It would be interesting to see if their metaphors remain consistent or evolve in some way.

## Conclusions

This study found that students used metaphorical language extensively in written responses to exam questions. This is significant for two reasons. First, this validates the methodological framework that was previously developed to analyze explicitly solicited analogies. In this study, we see that students use the same conceptual metaphors spontaneously in their writing, and so the same framework can be used to help identify the characteristics of energy in student responses. Additionally, this study helped to further refine the classification of conceptual metaphors. The 'energy as a substance that can change forms' metaphor was split into two subcategories: 'forms of energy' and 'transformation of energy' that reflect the difference in complexity of responses. The complications of distinguishing energy as an ingredient, product, or a substance to be stored were also revealed in this analysis.

Second, the students were writing about the role of energy in different scientific contexts. In these particular contexts, students found it useful to employ multiple coherent metaphors to explain the role of energy in these systems. While the set of conceptual metaphors identified previously may not be complete, they do have broader applicability than just traditional science courses. This is interesting because these students did not have a strong disciplinary basis on which to draw;

they had little exposure to the metaphors used in traditional scientific discourse. This suggests that the conceptual metaphors represent a conceptualization of energy that goes beyond the disciplinary structures and into everyday understandings, and that the disciplinary structures are not divorced from everyday understandings.

## Disclosure Statement

No potential conflict of interest was reported by the author.

## Notes

1. Note that some researchers use the term analogy to refer to the mapping between cognitive domains, but I am using it here in the literary sense to describe the actual language used by students.
2. An isolated system is one that is isolated from the surrounding environment. An open system is one in which energy can be transferred to and from the surrounding environment. By definition, energy is not conserved in open systems.
3. Usable energy refers to energy that can do work in a system, as opposed to energy dissipated to the environment (and therefore lost from the system).
4. It is worth noting that textbooks do not necessarily reflect the ideals of the disciplines. However, they are a primary source of information for students, and an important resource for teachers. At lower levels, teachers may learn the content from the text, and at higher levels the professors use the text to help them translate the content into language the students can understand. For these reasons they merit critical examination.
5. Table 3 lists this as a common metaphor for chemical reactions, but this was primarily due to students employing the ingredient/product aspect of the metaphor and *not* the energy storage metaphor. See Lancor (2014b) for details.

## References

American Association for the Advancement of Science. (2008). *Project 2061: Benchmarks online*. Retrieved from http://www.project2061.org/publications/bsl/online/index.php?chapter=4

Amin, T. G. (2009). Conceptual metaphor meets conceptual change. *Human Development, 52*(3), 165–197.

Anderson, B. (1990). Pupil's conceptions of matter and its transformations (age 12–16). *Studies in Science Education, 18*(1), 53–85.

Aubusson, P. J., Harrison, A. G., & Ritchie, S. M. (2006). Metaphor and analogy. In P. J. Aubusson, A. G. Harrison, & S. M. Ritchie (Eds.), *Metaphor and analogy in science education* (pp. 1–9). Dordrecht: Springer.

Barak, J., Sheva, B., Gorodetsky, M., & Gurion, B. (1999). As 'process' as it can get: students' understanding of biological processes. *International Journal of Science Education, 21*(12), 1281–1292.

Bell, B., & Cowie, B. (2001). The characteristics of formative assessment in science education. *Science Education, 85*(5), 536–553.

Black, P., & Wiliam, D. (1998). Assessment and classroom learning. *Assessment in Education, 5*(1), 7–74.

Boo, H. K. (1998). Students understandings of chemical bonds and the energetics of chemical reactions. *Journal of Research in Science Teaching, 35*(5), 569–581.

Campbell, N. A., & Reece, J. B. (2002). *Biology* (6th ed). San Francisco, CA: Pearson.

Chang, R. (1998). *Chemistry* (6th ed.). Boston, MA: WCB McGraw-Hill.

Collins, A., & Gentner, D. (1987). *How people construct mental models*. Retrieved June 23, 2010, from http://groups.psych.northwestern.edu/gentner/papers/CollinsGentner87.pdf

Dreyfus, B. W., Geller, B. D., Gouvea, J., Sawtelle, V., Turpen, C., & Redish, E. (2014). Ontological metaphors for negative energy in an interdisciplinary context. *Physical Review Special Topics—Physics Education Research, 10*(2), 020108-1–020108-11.

Duit, R. (1987). Should energy be illustrated as something quasi-material? *International Journal of Science Education, 9*(2), 139–145.

Duit, R. (1991). On the role of analogies and metaphors in learning science. *Science Education, 75*(6), 649–672.

Duit, R., & Haeussler, P. (1994). Learning and teaching energy. In P. J. Fensham, R. F. Gunstone, & R. T. White (Eds.), *The content of science: A constructivist approach to its teaching and learning* (pp. 185–200). Bristol, PA: Falmer Press.

Falk, G., Herrmann, F., & Schmid, G. (1983). Energy forms or energy carriers? *American Journal of Physics, 51*(12), 1074–1077.

Feynman, R., Leighton, R., & Sands, M. (2006). *The Feynman lectures on physics*. San Francisco, CA: Pearson.

Freeman, S. (2007). *Biological science* (3rd ed.). San Francisco, CA: Benjamin Cummings.

Gayford, C. G. (1986). Some aspects of the problems of teaching about energy in school biology. *European Journal of Science Education, 8*(4), 443–450.

Harrison, A. G., & Coll, R. K. (Eds.) (2008). *Using analogies in middle and secondary science classrooms*. Thousand Oaks, CA: Corwin Press.

Hecht, E. (2007). Energy and change. *The Physics Teacher, 45*(2), 88–92.

Hestenes, D. (2006). *Notes on modeling theory*. Proceedings of the 2006 GIREP conference: Modeling in Physics and Physics Education. Retrieved August 31, 2011, from http://modeling.asu.edu/R&E/Notes_on_Modeling_Theory.pdf

Jewett, J. W. (2008a). Energy and the confused student I: Work. *The Physics Teacher, 46*(1), 38–43.

Jewett, J. W. (2008b). Energy and the confused student II: Systems. *The Physics Teacher, 46*(2), 81–86.

Jewett, J. W. (2008c). Energy and the confused student III: Language. *The Physics Teacher, 46*(3), 149–153.

Jewett, J. W. (2008d). Energy and the confused student IV: A global approach to energy. *The Physics Teacher, 46*(4), 210–217.

Kaper, W. & Goedhart, M. (2010). 'Forms of Energy', as an intermediary language on the road to thermodynamics? Part II. *International Journal of Science Education, 24*(2), 119–137.

Knight, R. D. (2007). *Physics for scientists and engineers: A strategic approach* (2nd ed.). University of Virginia, Boston: Addison-Wesley.

Lakoff, G., & Johnson, M. (1980). *Metaphors we live by*. Chicago, IL: University of Chicago Press.

Lakoff, G., & Johnson, M. (1999). *Philosophy in the flesh*. New York, NY: Basic Books.

Lancor, R. A. (2014a). Using metaphor theory to examine conceptions of energy in biology, chemistry, and physics. *Science & Education, 23*(6), 1245–1267.

Lancor, R. A. (2014b). Using student-generated analogies to investigate conceptions of energy: A multidisciplinary study. *International Journal of Science Education, 36*(1), 1–23.

Leach, J., Driver, R., Scott, P., & Wood-Robinson, C. (1996). Children's ideas about ecology 2; ideas found in children aged 5–16 about the cycling of matter. *International Journal of Science Education, 18*(1), 19–34.

Lee, H., & Liu, O. L. (2010). Assessing learning progressions of energy concepts across middle school grades: The knowledge integration perspective. *Science Education, 94*(4), 665–688.

Lemke, J. (1997). *Teaching all the languages of science: Words, symbols, images, and actions*. Retrieved from http://www-personal.umich.edu/~jaylemke/papers/barcelon.htm

Lin, C., & Hu, R. (2003). Students' understanding of energy flow and matter cycling in the context of the food chain, photosynthesis, and respiration. *International Journal of Science Education*, 25(12), 1529–1544.

Nordine, J., Krajcik, J., & Fortus, D. (2011). Transforming energy instruction in middle school to support integrating understanding and future learning. *Science Education*, 95(4), 670–699.

Novik, S. (1976). No energy storage in chemical bonds. *Journal of Biological Education*, 10(3), 116–118.

Pintó, R., Couso, D., & Gutierrez, R. (2005). Using research on teachers' transformations of innovations to inform teacher education. The case of energy degradation. *Science Education*, 89(1), 38–55.

Posner, G. J., Strike, K. A., Hewson, P. W., & Gertzog, W. A. (1982). Accommodation of a scientific conception: Toward a theory of conceptual change. *Science Education*, 66(2), 211–227.

Scherr, R. E., Close, H. G., McKagan, S. B., & Vokos, S. (2012). Representing energy. I. Representing a substance ontology for energy. *Physical Review Special Topics—Physics Education Research*, 8(2), 020114-1–020114-11.

Sefton, I. M. (2004). *Understanding energy*. Retrieved from http://science.uniserve.edu.au/school/curric/stage6/phys/stw2004/sefton1.pdf

Serway, R., Faughn, J., & Vuille, C. (2006). *College physics* (7th ed.). Belmont, CA: Brooks/Cole.

Swackhamer, G. (2005). *Cognitive resources for understanding energy*. Retrieved from http://modeling.la.asu.edu/modeling/CognitiveResources-Energy.pdf

Taber, K. (1989). Energy—By many other names. *School Science Review*, 70(252), 57–62.

Tamir, P. (1998). Assessment and evaluation in science education: Opportunities to learn and outcomes. In B. Fraser, & K. Tobin (Eds.), *International handbook of science education* (pp. 761–790). Dordrecht: Kluwer Academic Publishers.

Trumper, R. (1990). Being constructive: An alternative approach to the teaching of the energy concept—Part one. *International Journal of Science Education*, 12(4), 343–354.

Watts, D. M. (1983). Some alternative views of energy. *Physics Education*, 18(5), 213–217.

Zurcher, U. (2008). Human food consumption: A primer on nonequilibrium thermodynamics for college physics. *European Journal of Physics*, 29(6), 1183–1190.

# Understanding Starts in the Mesocosm: Conceptual metaphor as a framework for external representations in science teaching

Kai Niebert[a] and Harald Gropengiesser[b]

[a]Science and Sustainability Education, University of Zurich, Zurich, Switzerland;
[b]Institute for Science Education, Leibniz University Hannover, Hannover, Germany

In recent years, researchers have become aware of the experiential grounding of scientific thought. Accordingly, research has shown that metaphorical mappings between experience-based source domains and abstract target domains are omnipresent in everyday and scientific language. The theory of conceptual metaphor explains these findings based on the assumption that understanding is embodied. Embodied understanding arises from recurrent bodily and social experience with our environment. As our perception is adapted to a medium-scale dimension, our embodied conceptions originate from this mesocosmic scale. With respect to this epistemological principle, we distinguish between micro-, meso- and macrocosmic phenomena. We use these insights to analyse how external representations of phenomena in the micro- and macrocosm can foster learning when they (a) address the students' learning demand by affording a mesocosmic experience or (b) assist reflection on embodied conceptions by representing their image schematic structure. We base our considerations on empirical evidence from teaching experiments on phenomena from the microcosm (microbial growth and signal conduction in neurons) and the macrocosm (greenhouse effect and carbon cycle). We discuss how the theory of conceptual metaphor can inform the development of external representations.

## Introduction

In the teaching of science, students' conceptions have come into the focus of science educators during the last four decades (overview in Duit, 2009). This research draws primarily on the perspective that was expressed by the educator Adolph Diesterweg as early as 1835 (p. 131): 'Without knowing about the students' viewpoints no proper instruction is possible'. Over the years, the research on students' conceptions has been embedded in various theoretical frameworks with epistemological, ontological and affective orientations (Duit & Treagust, 2003). Within the framework of educational reconstruction (Duit, Gropengiesser, Kattmann, & Komorek, 2012), we published a number of interview studies and teaching experiments on various topics. To interpret the nature of the students' conceptions, we referred to the theory of conceptual metaphor of Lakoff and Johnson (1980), in which they state that all of our knowledge draws on bodily and cultural experience. We have adopted this perspective of 'embodied conceptions' (Lakoff, 1990) to analyse and categorise students' conceptions on topics like cell biology (Riemeier & Gropengiesser, 2008), climate change (Niebert & Gropengiesser, 2014), physiology (Gropengiesser, 1997) and on scientific processes like experimentation (Niebert, 2007). When analysing these conceptions, we found that often very basal experiences—like those of containers, moving on a straight path or in a circle, being in or losing balance, or sharing and dividing things—constitute our basic understanding of scientific phenomena. These experiences are conceptualized in terms of image schemas, abstractions from sensorimotor experience (Johnson, 1987). For adequate understanding of science, both the selection of an embodied source and also the way this source is mapped to the phenomenon to be understood play major roles (Niebert, Marsch, & Treagust, 2012).

In these prior studies, we used the notion of embodied conceptions as a lens to analyse conceptions. The study at hand here is motivated by our interest to find out if, and how, an analysis of students' and scientists' embodied conceptions can not only help science educators to understand the origin of these conceptions but also inform the way we teach science. Therefore, in this paper, we are widening our perspective from how conceptions can be analysed regarding their embodied basis to how we can use embodied conceptions to teach science and develop learning activities.

*Conceptual Metaphor as a Theory of Understanding*

A growing number of researchers in cognitive science have discussed evidence that grant the body a central role in shaping the mind (Fauconnier & Turner, 2002; Johnson, 1987; Lakoff, 1990; Lakoff & Johnson, 1980; Rohrer, 2001, 2005). The various frameworks used in this research can be summed up under the umbrella term of embodied cognition, which refers to the view that cognitive processes are rooted in the body's interactions with its physical and cultural environment. This position houses a number of claims (for an overview see Wilson, 2002), from which we will mainly focus on one in our analyses: cognition is body-based.

Consider the following constructs where scientists make use of everyday experience to explain their theories. Robert Hooke was the first to denote the *cell* using the term 'cell' when an image of a piece of cork under his microscope reminded him of the small rooms, or cells, occupied by monks in monasteries. Kepler developed his concept of planetary motion by comparison with a clock. Huygens used water waves to theorise that light is wavelike. Arrhenius described the greenhouse effect by referring to his experience with hot pots. In ever new variations, scientists employ experiences from everyday life to understand scientific phenomena. Semino (2008) has pointed out that those metaphorical constructs are not only used with a pedagogical purpose, but also in many cases have a theory-constitutive function as well. But why do even scientists have to rely on bodily experiences to construct and explain their scientific ideas?

In the 1980s, linguists began exploring how understanding abstract concepts are regularly based on bodily concepts through metaphor (Lakoff & Johnson, 1980). An important finding from their research is that many concepts are not understood literally but metaphorically in terms of another domain of knowledge. Lakoff and Johnson argued that their findings of the omnipresence of metaphors were not only a linguistic phenomenon but also reflect 'general principles of understanding' (1980, p. 116). The referral to everyday experience and the use of metaphors are not just a matter of figurative language but are of a conceptual nature. This notion led to the development of the theory of conceptual metaphor.

Each conceptual metaphor has the same mode of operation: the structure of the (embodied) source domain is metaphorically projected to the target domain to achieve understanding. The embodied conceptions in the source domain provide an inference pattern to reason about the target domain. When the inferential logic is carried over from the source domain to the target domain, we regard that as a conceptual metaphor. A conceptual metaphor can be defined as a unidirectional mapping of entities from a concrete conceptual domain to what is usually more abstract conceptual domain. The ability of metaphorical thought makes abstract scientific theorising generally possible (Lakoff & Núñez, 2000).

The embodied sources of metaphors are often what Lakoff and Johnson (1980) call 'image schemas'. These image schemata—like the start-path-goal schema, an up-down schema or a front-back schema (Lakoff, 1990) —arise from recurrent experience, i.e. the interactions of our sensorimotor system with the environment. For example, the container schema emerges from our experience with our bodies as three-dimensional containers into which we put certain things such as food, water or air and out of which other things such as air, blood and waste emerge (Johnson, 1987). Image schemata give coherence and structure to our conceptions and are directly meaningful for orientation in our physical and social environment. We use the structures of these image schemata to understand abstract ideas that are not directly grounded in experience.

*Embodied Conceptions from an Epistemological Perspective*

The theory of conceptual metaphor helps us explain why we have problems understanding science concepts such as the theory of relativity, the theory of evolution and the cell theory. One line of reasoning points to the abstract nature of these theoretical notions and the necessity of imaginative thought (Lakoff, 1990). Closely related but more basic is the argument for the lack of direct experience of these processes.

Vollmer (1984) argues that our sensory system is not able to perceive or process phenomena like these. Based on approaches from evolutionary epistemology, he argues that the principles that underlie our cognitive processes were developed during human evolution. Our sensory and cognitive systems fit—at least partially—to the world we live in because they have emerged in a process of adaptation to the world. Vollmer calls the parts of the real world to which man has adapted his perception, experience and actions the mesocosm. It is a world of middle dimensions: medium distances and times and low velocities and forces. It extends from a blink to a lifetime, from light as a feather to heavy as an elephant, from a hair's breadth to the horizon and so forth (Table 1). These dimensions explicitly refer to human sensory abilities and are perceivable and tangible. The mesocosm is 'that section of the real world we cope with in perceiving and acting, sensually and motorically [ . . . ]' (Vollmer, 1984, p. 87).

Whereas perception and experience in general are primarily influenced by the mesocosm, scientific evidence and theories often exceed the mesocosm; macrocosmic structures such as the biosphere and the solar system are not part of the mesocosm because our cognitive system is not adapted to these dimensions. The same holds for microcosmic entities such as cells or structures such as molecules. To extend the mesocosmic boundaries, scientists often rely on complex technology and inquiry to open phenomena in the micro- and macrocosm to experience. In the macrocosm and microcosm, we encounter entities that are imperceptible, at least in our everyday experience.

These epistemological considerations support the theory of conceptual metaphor and explain the findings of researchers in science education investigating topics such as entropy (Amin, Jeppsson, Haglund, & Strömdahl, 2012; Jeppsson, Haglund, Amin, & Strömdahl, 2013), energy (Amin, 2009), thermodynamics (Fuchs, 2007), different mathematical concepts (Lakoff & Núñez, 2000), climate change (Niebert & Gropengiesser, 2013b), glacial movement (Felzmann, 2014) or physiological aspects like 'seeing' (Gropengiesser, 1997) that scientific concepts are regularly understood by using conceptual metaphors.

Some of these authors propose to make these findings applicable to science teaching. One of the most concrete proposals was made by Amin (2009, p. 192), who stated that the 'tools of conceptual metaphor can also support the design of instructional representational tools'. But an analysis of how embodied conceptions can inform the development of external representations is still a desideratum.

Table 1. Dimensions and boundaries of the mesocosm (cf. Vollmer, 1984)

| | Lower boundary | Upper boundary |
|---|---|---|
| Time | seconds (e.g. heartbeat) | decades (e.g. lifetime) |
| Range | millimetre (e.g. hair: 0.1 mm) | kilometre (e.g. daytrip: 30 km, horizon: 20 km) |
| Speed | $v = 0$ (e.g. rest) | $v = 10$ m/s (e.g. runner and preying bird) |
| Acceleration | $a = 0$ (e.g. steady motion) | $a = 10$ m/s$^2$ (runner and free fall) |
| Weight | gram (e.g. ping-pong ball) | ton (e.g. tree, animal and rock) |
| Temperature | $0°C$ (e.g. freezing point) | $100°C$ (e.g. boiling of water) |

*External Representations in Science Education*

Teaching always involves some way of representing information about scientific concepts and the phenomena to which they relate. But what a representation is, is difficult to define (Gilbert & Treagust, 2009). In accordance with science education literature (e.g. Gilbert & Treagust, 2009; Tsui & Treagust 2013), we use the term 'representation' to refer primarily to constructs of phenomena that come in the form of models, analogies, figures, diagrams, written or spoken text and so on. Even experiments and observations often serve as representations, as they represent, 'by example', some concept or phenomenon.

From a constructivist perspective, external representations can help students make sense of complex phenomena by constructing their own conceptions and avoiding alternative conceptions. Furthermore, research shows that learners can benefit from learning with more than one external representation (Tsui & Treagust, 2013). However, in an analysis of learning with external representations, Van Someren, Reimann, Boshuizen, and de Jong (1998) argue that students are often unable to make connections between different external representations or between an external representation and their prior conceptions. Moreover, there is evidence that in more than a few cases, representations used with an instructional purpose are not adequately understood by students (Harrison & de Jong, 2005), nor are they understood in the anticipated way (Harrison & Treagust, 2006). These empirical findings give evidence for what every science teacher knows from his own practise: some external representations are more effective than others.

In chemistry education, Johnstone's (1982) level-based description of external representations has become a dominant framework. Johnstone proposed that chemical knowledge is generated and communicated at three different levels: the symbolic, submicro and macro levels:

- External representations on the macro level[1] describe learning activities focussing on empirical properties of chemicals that are perceptible (e.g. mass, density, concentration, pH and temperature).
- Submicroscopic external representations are models or diagrams to explain macroscopic phenomena. These models represent entities that are too small to be perceived such as atoms, molecules or ions.
- Symbolic external representations involve conventions to represent atoms or molecules, signs to represent electrical charge, equations to show the conservation of matter during a reaction and so on.

This triplet of external representations has served as a framework for many studies and inspired the work of chemistry teachers and researchers as well (Gilbert & Treagust, 2009). While the triplet relationship has become a key model for chemical education, there is considerable evidence that students have problems in using the triplet relationship for understanding chemistry, as often no suitable experience is provided to the students (Nelson, 2002). Students are uncertain about how to connect the experience to their prior knowledge (Hodson, 1990), or they have difficulties translating between the macro and the submicro levels of representation (Davidowitz & Chittleborough, 2009). Moreover, Johnstone's representation triplet has a limited scope when it comes to teaching concepts from life and earth sciences, as knowledge in these domains extends to multiple entities—e.g. from evolution as an overall framework to different levels of explanation (molecule, cell, organism, population, biosphere, etc.).

To address these problems, Tsui and Treagust (2013) developed a three-dimensional model for teaching and learning biology with external representations. Within this model they argue that learning can take place by translating across:

- modes of representations with increasing abstraction from real-life worldly objects and actions to more abstract graphs, equations or verbal descriptions;
- levels of representation from the symbolic level (explanatory mechanisms), the submicro level (molecules), the micro level (organelles and cells) and the macro level (tissues and organs);
- content areas of biology, for example, connecting the ecological aspects (i.e. the carbon cycle) with physiological activities (i.e. photosynthesis and respiration) as discussed in our case three.

This model and the findings on students' difficulties in working with external representations show why learning the life and earth sciences is challenging: Understanding these sciences demands moving mentally in structurally and functionally related content areas (like evolution, homeostasis, energy, etc.) and skipping back and forth between the different levels of familiar and concrete vs. unfamiliar and abstract representations.

### External Representations and Conceptual Metaphors

The notion that understanding is embodied, even when it comes to concepts far from the mesocosm, should have implications for how we conceive external representations to teach science. Often a scientific concept is given an abstract definition or characterisation, which is viewed as the learning objective. This may take the form of a verbal definition, a formula, a model, a concept map and so on. In their discussion of the implications of the conceptual metaphor perspective for mathematics education, Núñez, Edwards, and Matos (1999) noted that teaching solely with abstract characterisation of concepts misses the reality of their roots in embodied conceptions. Taking the importance of embodied conceptions into account, experiential resources can support the design of external representations (e.g. Amin, 2009). Embodied conceptions can be accounted for by designing external representations that embody the abstract relations among the target concepts.

### Research Questions

The purpose of this study is to find out how students' and scientists' embodied conceptions can serve as a framework to support developing external representations of micro- and macrocosmic phenomena. Therefore, we are dealing with two research questions in our paper:

- On which embodied conceptions do students and scientists draw to understand selected phenomena from the micro- and macrocosm?
- How can embodied conceptions inform the design of external representations of selected micro- and macrocosmic phenomena?

To serve this purpose, we draw on previous studies on students' conceptions and new empirical data to analyse students' and scientists' embodied conceptions. Based on these findings, we present data from teaching experiments on microscopic phenomena (microbial growth and signal conduction) and macroscopic phenomena (greenhouse effect and carbon cycle) in which we developed and probed external representations that engage students' embodied conceptions.

## Research Design and Methods

The empirical data that we refer to in the analysis reported in this paper were collected as part of a larger project carried out within the model of educational reconstruction. The model of educational reconstruction is a widely used research programme that was developed to improve content specific learning and teaching (Duit et al., 2012; Kattmann, Duit, & Gropengiesser, 1998; Niebert & Gropengiesser, 2013a). As a research programme, the model of educational reconstruction identifies and interrelates three relevant research tasks of subject matter education: (a) critical analysis of science content, (b) investigation into students' perspectives and (c) analysis, design and evaluation of learning environments. Using the model of educational reconstruction as a research design, we conducted several teaching experiments (Komorek & Duit, 2004; Steffe & Thompson, 2000) in order to analyse students' conceptions of different phenomena and to evaluate their conceptual development when interacting with external representations.

In this study, we report the results of teaching experiments with 118 students on concepts from microcosm (cell division and neurobiology) and macrocosm (greenhouse effect and carbon cycle; see Table 2). Each teaching experiment starts with a short interview investigating students' conceptions. This interview is followed by a sequence of teaching episodes. The teaching experiments lasted 45–90 min and were conducted with dyads or triads of students on the premises of the University of Hannover and the Leuphana University Lueneburg.

The external representations probed in our teaching experiments were based on data on students' conceptions from prior studies on cell division (Riemeier & Gropengiesser, 2008), signal conduction (Fichtner, 2013), the greenhouse effect (Niebert & Gropengiesser, 2014) and the carbon cycle (Niebert & Gropengiesser, 2013b). In the study at hand here, we reanalyse these data with the aim of identifying the students' embodied conceptions so as to inform the design of external representations. Following the principles of the model of educational reconstruction, we also analysed the embodied conceptions of scientists from textbooks and research reports (Table 2). To analyse the mesocosmic experience guiding students' and scientists' conceptions, we conducted a metaphor analysis (Table 3). We present the results at the level of conceptual metaphors

Table 2.   Sources of data presented in this study

| Topic | Students' conceptions | Scientists' conceptions |
|---|---|---|
| Microbial growth | 48 secondary school students (16 triads), (15–16 yrs.) | Campbell et al. (2008) |
| Signal conduction | 13 undergraduate students (5 dyads, 1 triad; 19–24 yrs.) | Campbell et al. (2008) |
| Carbon cycle | 39 secondary school students (9 triads, 6 dyads, 17–19 yrs.) | IPCC (2013) |
| Greenhouse effect | 18 secondary school students (2 triads, 6 dyads, (17–19 yrs.) | IPCC (2013) |

as *Target Is Source* (Lakoff & Johnson, 1980). This level seems appropriate as (a) the nearly unlimited variety of linguistic expressions that can describe an aspect can be categorised into a limited number of conceptual metaphors (Niebert et al., 2012; Schmitt, 2005) and (b) a conceptual metaphor clearly makes visible the mesocosmic experience that guides understanding. We take this analytical level for our design of external representations informed by conceptual metaphor theory.

Table 3.   Steps used in metaphor analysis

| Steps | Examples, metaphors are marked in italics |
|---|---|
| 1. *Identifying Metaphors*<br>We identified all metaphors in the material and chose the metaphors that were crucial for understanding neurobiology. | '[ . . . ] Schwann cells *wrap themselves around* axons, *forming layers* of myelin'<br>'*In* a myelinated axon, the *depolarizing current during an action potential* at one *node* of Ranvier *spreads along* the *interior* of the axon to the next *node* [ . . . ], where it *reinitiates itself*. Thus, the action potential *jumps from node to node* as it *travels along* the axon'. (Campbell et al., 2008, p. 1056) |
| 2. *Finding Conceptual Metaphors*<br>We arranged all metaphors with the same target and source domains. | • *Wrap themselves around, forming layers*: Myelin Is Forming Layers;<br>• *In* a axon, *interior* of the axon: Axon Is Container;<br>• The action potential *jumps, travels* and *reinitiates itself*: Action Potential Is Travelling Agent; Action Potential Is Jumping Agent Action Potential Is Moving Agent |
| 3. *Interpreting Conceptual Metaphors*<br>We described the metaphorical patterns used by students and scientists guided by embodied cognition. The conceptual metaphors described in this paper are denoted by capitalised letters (Target Is Source) | When using terms like *jumping, travelling* and *reinitiating,* the process of an action potential is reified as an active agent. This agent travels through the axon (which is imagined to be a long drawn out container with a path inside), where every node of Ranvier is a start and a goal of saltatory signal conduction |

For our metaphor analysis, we analysed conceptual metaphors in the transcripts of the interviews, in science textbooks and research reports by discerning terms or sequences that have, or may have, more than one meaning. Our adaption of the metaphor analysis is presented by way of example in Table 3 with the example of saltatory signal conduction (Campbell et al., 2008, p. 1056).

Based on the analysis of students' embodied conceptions, we developed external representations that relate to these conceptions. To base these external representations on embodied conceptions, we first defined the students' learning demand by comparing the embodied conceptions scientists and students hold. We then evaluated the effects of the external representations on students' conceptual development in teaching experiments. The external representations we probed are presented and described in the results section. To analyse the students' conceptual development we conducted a qualitative content analysis (Mayring, 2002) in which: (1) we transcribed the students' interactions during the teaching experiments and edited the texts to improve readability; (2) we arranged the statements by content and (3) we interpreted the statements about the underlying conceptions. In addition, we conducted a metaphor analysis of the students' conversations during the teaching experiments to compare the students' conceptual metaphors before and while interacting with the external representations.

## Results

In this section, we illustrate how the theory of conceptual metaphor can inform the analysis of student conceptions and the design of external representations. Therefore, we present teaching experiments on microbiology, neurobiology, the carbon cycle and the greenhouse effect. The external representations applied in these teaching experiments were developed based on analyses of students' and scientists' embodied conceptions. Therefore, every subsection starts by explicating these embodied conceptions and analysing the learning demand.

### External Representations of Microbial Growth

In a previous study on students' conceptions of the concept of growth, Riemeier and Gropengiesser (2008) found that 7th grade students are able to explain the growth of onion roots by referring to the phrase cell division: 'The growth happens by cell division'. From a scientific perspective, the students refer to an adequate scientific concept. But a deeper analysis of their understanding of the concept 'cell division' reveals a conceptual misunderstanding of the term *division*: Asked to explain their conceptions of cell division, a typical student's answer was, 'Division can lead to multiplication of cells [makes a cut in two with her hands]'. One student outlined her conception in a drawing (see Figure 1).

For the purposes of this study, our reanalysis of the embodied conceptions forming students' understanding of cell division reveals that the students adhere to a *division image schema* that is combined with a *part–whole schema*. In these schemata, division is conceptualised as resulting in (a) more single parts than the whole and (b) smaller

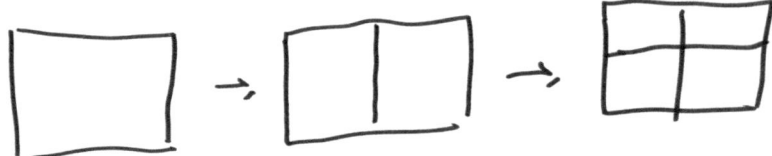

Figure 1.   Drawing of cell division by a student aged 15 years

parts than the whole. A division of a whole (cell) results in two parts (cells). But these cells are not identical to the whole; they are half the size of the original part (cell).

To conceptualise growth, the students think of division, exclusively, as becoming more cells. They construct their understanding of cell division and thus, the growth of organisms, based on the conceptual metaphor *Growth Is Division*. In this conceptual metaphor, students construe division as *becoming more and more* cells which is rashly equated to growth. The second element of the division schema *becoming smaller* is not mapped to the target domain. Thus, the students get the idea that *more cells* suffice to accomplish the growth of organisms. To scientists, cell division (mitosis) implies the division of cells accompanied by the growth of cells (Campbell et al., 2008). They construct the conceptual metaphor *Growth Is Division and Enlargement*.

In our interpretation, the mapping of the elements of the division schema (Dividing Is Becoming More and Dividing Is Becoming Smaller) and the part–whole image schema causes obstacles to understanding the concept of cell division. The students' learning demand requires a meticulous mapping of the elements of the schemata. Further analysis of the students' learning demand based on their conceptual metaphors reveals that they have to understand that *cell division* is based on the concept of *enlargement* as well.

We developed two external representations to support students' conceptual development in the topic of cell division and probed them in a Ph.D. study on students' understanding of bacteria (Schneeweiss, 2008). This context is comparable to the setting of Riemeier and Gropengiesser (2008) since growth of bacteria is based on the same principles (mitosis) as the growth of onion roots.

- The external representation 'colony growth' aimed to project cell division to a mesocosmic level: While one bacterium (one cell) is part of the microcosm and invisible to the naked eye, a colony of bacteria (cells) is part of the mesocosm. Therefore, some bacteria were incubated in a petri dish with agar-agar for 24 hours. After incubation, the students were able to see the grown bacteria colonies.
- The external representation 'tearing paper' aimed to encourage students to reflect on the use of the division schema and the part–whole schema: The students were asked to tear a sheet of paper into squares and subsequently compare this process to cell division. When tearing paper, the mass of the whole remains the same, the number of parts doubles, but the parts are smaller. This external representation aimed to offer an opportunity for reflecting on the contradiction between the reduction of the size of one cell by division and the growth of an organism or a colony of bacteria.

Initially, asked to explain the growth of a bacteria colony, these students used the same conceptual metaphor as the students in the study of Riemeier and Gropengiesser: *Growth Is Division*:

> Kim: The colony grows because the cells divide at the membrane. Then there are two bacteria and so on, until a colony becomes visible.

The student Kim imagines that an increase in the number of bacteria is sufficient to explain the growth of a bacteria colony. But division by itself cannot explain an increase in biomass, which is a precondition for the colony to become visible to the naked eye. To address this use of the conceptual metaphor *Growth Is Division* the external representation 'tearing paper' is introduced, and Kim argues while working with the external representation:

> Kim: The cells must divide and grow again to the size of the mother cell. If the bacteria just divide you cannot see them. Then we would have a lot of small bacteria, but the size of all bacteria would be as small as the one before.

Kim maps her experience with tearing paper to the growth of a bacteria colony. She recognises that dividing has two meanings: *becoming more* and *becoming smaller*. By reflecting in this way, she infers that growth must be a result of a regrowth of the smaller parts to the size of the former cell. A mapping of the part–whole image schema to cell division becomes obvious: The parts (daughter cells) have to regrow to a whole (size of the mother cell) to form a visible colony. Reflecting on the part–whole image schema and the use of the conceptual metaphor *Growth Is Division* initiated a conceptual development to resemble the conceptual metaphor used by scientists: *Growth Is Division and Enlargement*.

In another case, the external representation 'colony growth' was sufficient to initiate a conceptual development. Another student, named Tom, explains the occurrence of a visible bacteria colony based on nutrition of bacteria:

> Tom: The bacteria form a colony, because they take nutrients from the agar. These makes the bacteria grow and they divide and form a colony.

Tom constructs a conceptual metaphor based on another everyday experience: *Growth Is Enlargement by Nutrition*. He construes the agar as nutrition for the bacteria, which enables them to grow. After they have grown, they can divide again. This explains the visible increase in biomass. His argumentation is based on two conceptual metaphors: *Growth Is Enlargement by Nutrition* and *Growth Is Division and Enlargement*.

*External Representations of Saltatory Signal Conduction*

The conduction speed in axons of vertebrates that are insulated by myelin sheaths is considerably faster than in unmyelinated axons. The insulation is interrupted by nodes of Ranvier where the depolarising current triggers an action potential. The action potential at one node will depolarise the neighbouring node sufficiently. Fichtner (2013) found that teaching saltatory signal conduction solely based on figures of the anatomy of nerves and the physiology of action potentials by figures from a science textbook (Campbell et al., 2008) poses problems for students. Asked to describe the role of insulation by myelin for the conduction speed in neurons, the student Tina answered: 'I cannot imagine how myelin affects the traveling time of signals, it prevents ions from leaving the neuron. The distance a signal has to travel in the neuron is the same with or without myelin.'

Our reanalysis of these data shows that in her attempt to grasp the role of insulation in signal conduction, Tina imagines the nerve as being a container (*in* the neuron, *leaving* the neuron: *Neuron Is Container*). Within this conceptual metaphor myelin is the boundary of the container, which keeps the ions inside (*Myelin Is Boundary*). Asked how she conceptualises the conduction speed, Tina refers to the start-path-goal schema (*distance, travel: Neuron Is Path*). The start-path-goal schema consists of a start, an agent (signal) that moves in a certain direction and a goal (Lakoff & Johnson, 1999). In the mapping of this schema the signal is reified as a travelling agent that moves (*Signal Is Travelling Agent*). The travelling time (conduction speed) of a signal depends on the *range of the path* as the determining element (*Conduction Speed Is Depending on Range of Path*). For Tina the myelin is conceptualised within the container schema (Myelin Is Boundary) but not within the start-path-goal schema.

Scientists use the container *and* the start-path-goal schema to construe saltatory signal conduction, too (Campbell et al., 2008, p. 1056), see Table 3: *Neuron Is Container, Myelin Is Boundary, Neuron Is Path*. While using similar conceptual metaphors as Tina, the scientists refer to different elements of the start-path-goal schema to construct the idea of conduction speed: For them the conduction speed depends on the length of the path and the speed of the signal (*Conduction Speed Depends on Range of Path/Speed of Signal*). The speed of the signal is enhanced by myelin: Scientists construe the role of myelin not only within the container schema as a boundary, but also within the start-path-goal schema: Within the start-path-goal schema myelin has the role of a barrier. This barrier forces the signal to 'jump [ . . . over it . . . ] from node to node' (Campbell et al., 2008, p. 1056): *Myelin Is Barrier* and *Signal Is Jumping Agent*. When jumping, the speed of the signal is enhanced.

Students and scientists use similar conceptual metaphors to construct an understanding of signal conduction: *Neuron Is Container, Neuron Is Path, Signal Is Travelling Agent, Myelin Is Boundary*. Since these conceptual metaphors are the same in students' and scientists' thinking, obviously they are not sufficient to construct a scientific understanding of saltatory conduction. To construe saltatory signal conduction, scientists additionally use the conceptual metaphors *Myelin Is Barrier, Signal Is Jumping Agent* to construct the concept *Conduction Speed Depends on Speed of Signal*. As long as the students see signal conduction only as a journey of an agent at a certain speed, insulating of the path will not lead to the idea of faster travel. In our teaching experiments, we provided the external representation 'Toppling dominoes' with domino-bricks and drinking straws to model how the speed of signal conduction can be enhanced by bridging parts of the way (Figure 2). We aimed at bringing the use of the start-path-goal schema to the students' mind and introduce the straws (myelin) as a bridge and the impulse (signal) jumping over this bridge.

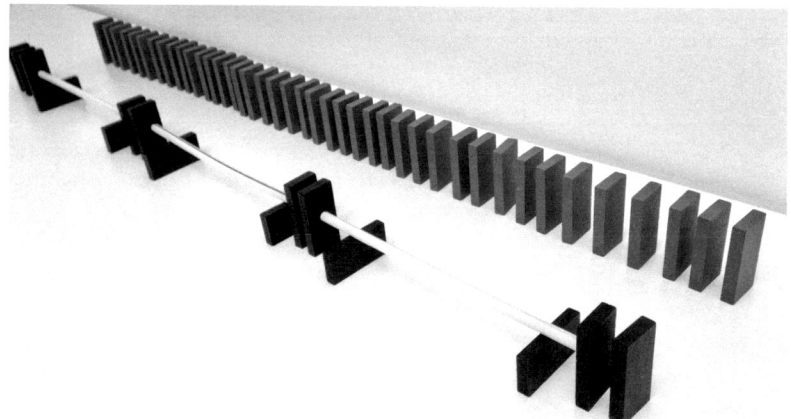

Figure 2.   External representation of signal conduction (a) and saltatory conduction (b) the self-propagating process of falling dominoes models a chain reaction, where one event starts off a chain of similar events. This step-by-step process works only if the next domino is toppled (a). A straw widens the step from one falling domino-brick to the next (b)

The following episode shows the discussion of two students (Amy and Ben) who were asked to model both axons (myelinated and unmyelinated) with domino bricks:

| | |
|---|---|
| Amy: | The conduction in the domino line with the straw was much quicker than in the line without straws. The straw kind-of bridges some dominoes. |
| Ben: | The straws are the myelin and the dominoes between the straws are these Ranvier rings. The straws and the myelin make the signal jump from one point to another. |
| Amy: | The time of the falling dominoes is the time the ion channels need to open. The fewer channels need to open, the faster the signal is transferred. |
| Ben: | Yes, it is like playing handball: You can quickly throw a ball or slowly hand over a ball from one player to another to get it goalwards. |

While working with the model, the students map the mesocosmic experiences with the model to their conceptions on signal conduction in microcosm in a set of conceptual metaphors: *Conduction Is Falling Dominoes, Straws Are Myelin, Dominoes Are Ion Channels*. In addition, they use these conceptual metaphors to express their newly developed conceptions: *Straw Is Bridge* and *Signal Is Jumping Agent*. These conceptual metaphors seem to be helpful for students to understand the process of signal conduction: When a signal jumps, it becomes faster; therefore they developed the idea *Conduction Speed Depends on Speed of Signal*.

At the end of the teaching experiment Ben mapped his newly developed conception to a situation from everyday life. He refers to the experience of giving and receiving a ball, i.e. in a game: Handing over the ball from one hand to another is slower than throwing a ball.

*External Representations of the Greenhouse Effect*

In a prior study, we analysed students' and scientists' conceptions of the greenhouse effect. Based on the conceptual metaphors they hold, we have shown that both students

Table 4. Thinking patterns on the causes of climate change

| | (a) Ozone hole | (b) Greenhouse layer | (c) Greenhouse atmosphere |
|---|---|---|---|
| Quote | *$CO_2$ makes a hole into the ozone-layer. More sunrays enter the atmosphere and the earth warms up* | *More $CO_2$ thickens the greenhouse layer. The layer captures sunrays in the atmosphere and it warms up.* | *$CO_2$ is evenly distributed in the atmosphere. More $CO_2$ shifts the radiative equilibrium.* |
| Conceptual metaphors | $CO_2$ Is Destroyer (of Boundary) Greenhouse Effect Is More Input | $CO_2$ Is Boundary (of Container) Greenhouse Effect Is Less Output | $CO_2$ Is Content (of Container) Greenhouse Effect Is Shifted Equilibrium |

and scientists construct the greenhouse effect with different mappings of the image schemata of containers and balances in the atmosphere (Niebert & Gropengiesser, 2014).

The reanalysis of our results reveals the embodied conceptions guiding students' and scientists' understanding: The thinking patterns outlined in Table 4 show that students (a and b) and scientists (c) map different structures of the container to the structures of the atmosphere resulting in different conceptual metaphors: where *$CO_2$ Is Destroyer (of Boundary)* by attacking the atmosphere (a), or *$CO_2$ Is (thickening) Boundary (of Container)* (b) or *$CO_2$ Is Content (of Container)* (c). In addition, different mappings of the balance schema can be found: *Greenhouse Effect Is More Input* (a), *Greenhouse Effect Is Less Output* (b) or *Greenhouse Effect Is Shifted Equilibrium* (c).

In the study at hand, we defined the students' learning demand as follows: (1) Students need the experience of $CO_2$ interacting with radiation and (2) Students need to reflect on the mapping of the balance schema to the greenhouse effect. To address these learning demands we developed the external representation 'Greenhouse effect' that brought the principles of the greenhouse effect to the mesocosm. The greenhouse effect was simulated in two big (2 l) glass beakers: one of the beakers is filled with air, the other with $CO_2$. Both beakers were irradiated using a 200 W lamp and the development of temperature was measured in the beakers. As the beakers had no lid on them, there is no upper boundary, nothing can be attacked; thus, the warming has to be due to another mechanism. This setup addressed the students' conceptual metaphor that *$CO_2$ Is Destroyer of Boundary* of the container in a more indirect way of disclosing the employed schema and to ask for pondering on its selective use for understanding the role of $CO_2$ in climate change. To clarify the position of $CO_2$ in the container atmosphere, we additionally measured the concentration of $CO_2$ at the bottom, in the middle and at the top of the beaker. The students Ann and Tim worked with this external representation:

Ann:    I thought that the ozone hole is responsible for warming. But it cannot be. I mean, we have no ozone layer here and it is warming anyway.

Tim:    That's what I told you, it's not the ozone hole: $CO_2$ captures the sun rays. [ ... ] I thought $CO_2$ forms a cloud. But this device shows that $CO_2$ is the same at the bottom and at the top. Does that mean this happens down here, where we live?

While working with the external representation of the greenhouse effect, the students discuss their conceptions in light of the evidence they found. Tim initially held the conception of the greenhouse effect using the conceptual metaphor *CO$_2$ Is Cloud* which is related to the conceptual metaphor *CO$_2$ Is Boundary* as both use the same spatial relations, even before working with the external representation. Ann initially stuck to the conceptual metaphor *CO$_2$ Is Destroyer (of Boundary)*. The external representation led her into a cognitive conflict and made her reject her initial conceptual metaphor. The external representation in itself gave no explanation for how $CO_2$ leads to warming. This explanation is generated by Tim who sticks to his initial conception. Tim even goes a bit further after interpreting the results of measuring the concentration of $CO_2$. For him, it seems to be hard to believe that the greenhouse effect happens around him.

The idea of *Greenhouse Effect Is Shifted Equilibrium* is necessary to understand the role of $CO_2$ in climate change. But a dynamic equilibrium is hard to understand because it combines, even in its simplest implementation, two embodied image schemata: a container and a balance. To understand the combined schemata is the learning demand in this case. We disclosed the combination of schemata directly with the external representation 'visualised balance schema' consisting of a beaker with a valve at the bottom, fed and drained by water. If the valve at the bottom was medium open, the inflow and outflow of water were constant. Students were asked to manipulate the in- and outflow of water and compare it to the amount of heat in the atmosphere. From the perspective of conceptual metaphor theory, this external representation focused on helping students to consolidate the source domain and to map it onto the shifting radiative equilibrium of the atmosphere.

After working with the external representation 'dynamic equilibrium' we prepared cards with written conceptions of global warming (wording and diagrams as presented in Table 4) without tagging them as every day or scientific. The container image schema was explicitly used. The following students' conversation was typical when arguing about the different conceptions:

Max:    The idea 'Warming By More Input' was what we initially thought. But it cannot be that way, because this would mean the ozone hole is involved—and it isn't. It's the $CO_2$ that stores the heat in the box, so it must be 'Warming By Less Output'.

Luke:   But if it is less output, more and more heat is captured in the atmosphere. The temperature would rise to infinity. I think it must be this »New Equilibrium«.

Max:    Yes, $CO_2$ stores heat and gives it away again. But the more $CO_2$ is in the atmosphere, the more heat is stored. [ ... ] It is like my pocket money: Until my birthday, I got 10 € a week—and spent everything. Now, I get 15 € every week, and there is nothing left at the end of the week, too. But now I can afford to go to the cinema in every week.

In their argumentation, the students Max and Luke connected the experience they made during the experiments to the schemata they used to understand global warming: At first, they rejected the conception *Greenhouse Effect Is More Input* and switched to *Greenhouse Effect Is Less Output*. This mechanism of capturing heat rays is a conception that is also presented in some textbooks. It is an oversimplified idea of the energy budget, which is not appropriate to achieve an adequate understanding. The experience of a dynamic equilibrium helped the students to construct the scientific idea of global warming.

At the end of the teaching experiment, Max applied the scientific conception and the image schemata he used to the everyday experience of getting and spending pocket money. Obviously, the experience and his reflecting on the container and the balance image schemata enabled him to discuss not only the scientific conception but also his everyday experiences. We argue that this works for him because events are often construed in relation to a container (Lakoff & Johnson, 1980), and the incoming and outgoing money per week is also interpreted as an equilibrium. Therefore, the students can use the same resources to understand the energy budget of the atmosphere as their own 'fiscal budget'.

### *External Representations of the Carbon Cycle*

In teaching experiments on the role of the carbon cycle in global warming we aimed to address a major problem reported by Sterman (2008). They probed students' ability to predict future $CO_2$ emissions and removal to mitigate global warming and informed students that today's $CO_2$ emissions are roughly twice the rate of net removal. Asked to predict the rate of $CO_2$ emissions and removal that is needed to stabilise the atmospheric $CO_2$ level, most students believed that stopping the growth of emissions stops the increase in $CO_2$ concentration. That vast majority of students (84%) asserted that the atmospheric $CO_2$ level would stabilise even though emissions exceed removal. This is in fact wrong—emissions and removal need to be the same to stabilise the $CO_2$ level.

To address this issue, we made use of a previous analysis of students' embodied conceptions of the carbon cycle (Niebert & Gropengiesser, 2013b). In this earlier study, we found that even if on a content level the conceptions of students differ widely from those of scientists, both draw on the same embodied conceptions: the image schemata of containers and balances form their conceptions which can be analysed from the conceptual metaphors they used: *$CO_2$ Is A Substance Stored, $CO_2$ Is A Substance Set Free, $CO_2$ Is A Substance Removed (from the Atmosphere)* (carbon pools are conceptualised as containers) or *$CO_2$ Is A Substance with Balanced Flow* or *$CO_2$ Is A Substance With Unbalanced Flow, Too Much $CO_2$ Disturbs Atmosphere* (carbon flows are conceptualised with the balance schema) (Niebert, 2007).

With these embodied conceptions in mind we used the setup of the external representation 'visualised balance schema' (see above) to foster students' understanding of the relation between the $CO_2$ emission/removal and the atmospheric $CO_2$ level. Before working with the external representation the students were asked to outline their conception in a graph: 'How do the $CO_2$ emissions and removals have to

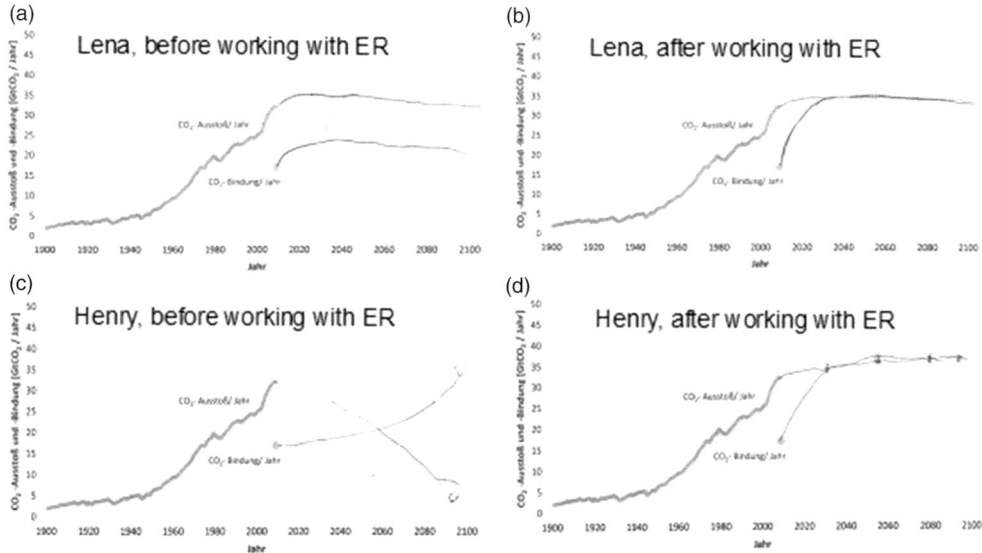

Figure 3. Lena's (a) and (b) and Henry's (c) and (d) graphs of $CO_2$ emissions and removal. Lena and Henry were asked to draw their conception on the development of $CO_2$ emission and removal to keep a constant concentration of $CO_2$ in the atmosphere; before (left) and after (right) working with the external representation 'dynamic equilibrium'

develop to keep a constant level of $CO_2$ in the atmosphere (i.e. limit global warming to 2°C).' This was the same task given by Sterman (2008).

The results (see Figure 3(a)) show that initially the student Lena had the same difficulties as reported by Sterman (2008): the emissions were stabilised but exceeded the removal. From the perspective of the balance image schema this conception is based on the idea *Constant $CO_2$ level Is Constant Input*. When working with the external representation Lena was asked to compare the amount of water in the beaker with the amount of $CO_2$ in the atmosphere:

> Lena: In global warming more water flows into the beaker than leaving it.
> Interviewer: Can you please map your findings to the atmosphere?
> Lena: To keep the temperature at a certain level, the input and output of water must be the same. Then the same amount of $CO_2$ must go into the atmosphere and leave it again.

In working with the external representation of the atmospheric $CO_2$ level, Lena starts by implicitly switching between arguing on the mesocosmic level of the beaker and the macrocosmic level of the atmosphere. She refers to the balance image schema to construct a conceptual metaphor to explain global warming: *Warming Is More Inflow*. This conceptual metaphor brings together the mesocosmic level of the water flow and the macrocosmic level of warming. She uses a related conceptual metaphor to construct an idea of how to keep the atmospheric temperature constant: *Stopping Warming Is Balancing Flows*. Here again she refers to the mesocosmic water flow as a source for understanding. Finally, this understanding is mapped by her to the atmosphere when she exchanges the source domain water flow to $CO_2$ flow (*Stopping Warming Is Balancing $CO_2$ Flows*). From the perspective of the balance schema she argues now with the conceptual metaphor *Constant $CO_2$ level Is Balancing Input and Output.*

After working with the ER, we asked Lena if she wants to redraw her initial diagram. The results presented in Figure 3(b) show that she is able to transfer the conceptual development initiated in working with the external representation to draw a revised and correct diagram.

Figure 3(c) shows the conception Henry initially held. In his conception, the removal has to overshoot the emission of $CO_2$ to keep a constant $CO_2$ level. His use of the balance schema reveals the conceptual metaphor *Constant $CO_2$ level Is More Output than Input.* When working with the external representation he argues as follows:

Henry:      The CO2 emissions are the inflowing water, the outflowing water determines the removal. The beaker is the atmosphere. [ ... ] We have a balance when input and output are the same. The less we emit, the less must be removed, to have a constant amount of CO2 in the atmosphere. [ ... ] If I emit less CO2 than is removed, then at some point there is no CO2 in the atmosphere and we get the next ice age.

In working with the external representation Henry maps his mesocosmic experience with the beaker to the macrocosmic phenomena in the atmosphere. In his comparison, he explicitly constructs the conceptual metaphors *$CO_2$ Is Water, Emission Is Input, Removal Is Output.* He starts his explanations in working with the external representation in the mesocosm ('balance if input and output are the same') and then switches to macrocosm ('the less we emit, the less must be removed') to construct his idea of a balanced $CO_2$ emission and removal. In the last section, he reflects on his initial conception, which is presented in the graph in Figure 3(c): If removal exceeds emission the atmosphere would cool down. After working with the external representation he is asked to redraw his diagram. The result in Figure 3(d) shows that he too is able to transfer his insights from working with the external representation to the diagram.

## Discussion

Our study was guided by the intention to find out how students' and scientists' embodied conceptions can serve as a framework to develop external representations of micro- and macrocosmic phenomena. In this section, we will discuss how conceptual metaphor theory can serve as a framework to identify the learning demand and to inform the design of external representations for teaching about micro- and macrocosmic phenomena.

### Conceptual Metaphors as a Level to Reveal the Learning Demand

For the design of external representations that are informed by conceptual metaphor theory a teacher needs to know about the students' learning demand. Therefore, research into students' conceptions is required—and these conceptions need to be analysed to reveal embodied conceptions. In order to develop an evidence-based

Table 5. Conceptual metaphors of Students and Scientists: Central pre-instructional conceptual metaphors of scientists and students discussed in this paper[a]

| Topic | Students' conceptual metaphors | Scientists' conceptual metaphors |
|---|---|---|
| Microbial growth | Dividing Is Becoming More Growth Is Division | Dividing Is Becoming More Dividing Is Becoming Smaller Growth Is Division and Enlargement |
| Signal conduction | Neuron Is Container, Myelin is Boundary, Neuron Is Path Conduction Speed Is Depending on Range of Path Signal Is Travelling Agent | Neuron Is Container, Myelin is Boundary Neuron Is Path, Myelin Is Barrier, Conduction Speed Is Depending on Range of Path Conduction Speed Is Depending on Speed of Signal Signal Is Jumping Agent |
| Greenhouse effect | Atmosphere Is Container $CO_2$ Is Boundary of Container/$CO_2$ Is Cloud, $CO_2$ Is Destroyer of Boundary Greenhouse Effect Is More Input Greenhouse Effect Is Less Output | Atmosphere Is Container $CO_2$ Is Content Greenhouse Effect Is Shifted Equilibrium |
| Carbon cycle | Carbon Pools Are Containers Constant $CO_2$ level Is Constant Input Constant $CO_2$ level Is Less Input than Output | Carbon Pools Are Containers Constant $CO_2$ level Is Balanced Input and Output |

[a]A full list of the conceptual metaphors we analysed is presented in Appendix.

formulation of students' learning demand, we additionally analysed scientists' conceptions. Both conceptions are discussed at the level of conceptual metaphors to have a basis for comparison. A summary of the conceptual metaphors of scientists and students analysed in this paper is presented in Table 5.

An analysis of the conceptual metaphors students and scientists use to construe the selected phenomena reveals the mesocosmic experience they draw on. Contrasting students' and scientists' conceptual metaphors is fruitful insofar as it provides a systematic perspective to categorise students' conceptions. Our analysis of the conceptual metaphors revealed that only a limited number of image schematic structures were employed in construing the four very different phenomena. This finding fits with the compilation of Mathewson (2005) who analysed the visual core of scientific understanding at the level of *master images*. Mathewson describes master images as being a condensed structure of the visual content of science—patterns, structures, objects and phenomena. In his analyses, he stated that scientific understanding is based on a limited list of 36 master images. These master images like containers, cycles, flows, paths, boundaries and so on. are conceptually closely related to the image schemata described by Johnson (1987).

Moreover, our analysis has shown that the number of image schemata used to understand the discussed phenomena is not only limited; in all of the analysed

Table 6.   Image schemata and sources of alternative conceptions to construe selected phenomena

| Topic | Image schema | Source of alternative conception |
|---|---|---|
| Microbial growth | Division schema Part–whole schema | Just parts of the division schema are mapped to construe cell division |
| Signal conduction | Container schema Start-path-goal schema | Construct Myelin in container schema but not in start-path-goal schema Construe conduction speed by range of path but not by speed of agent |
| Greenhouse effect | Container schema Balance schema | Not adequate mapping of $CO_2$ in the container schema Solely focussing on input or output in balance schema to the atmosphere |
| Carbon cycle | Container schema Balance schema | Solely focussing on input or output in balance schema to the carbon budget |

cases students and scientists referred to the *same* image schemata. However, although they draw on the same image schemata for understanding a phenomenon the constructed alternative conceptions are very different from the scientific ones. These can be traced back to selective mappings when constructing the conceptual metaphors (Table 6). Clearly, besides the selection of the source domain, the mapping of the different elements of an image schema is crucial for scientific understanding. This supports the hypothesis formulated by Amin (2009, p. 193) that 'learning the conventional mappings underlying the metaphoric expressions in scientific discourse constitutes an underappreciated obstacle to achieving conceptual change'.

*The Literal and Metaphorical Use of Image Schematic Structures*

In our analysis we interpreted conceptions related to the greenhouse effect or the carbon cycle based on the conceptual metaphor *Atmosphere Is Container*. But is this really a metaphorical construal of the atmosphere? The atmosphere is located in a spatial domain, so are not terms such as *emission, in the atmosphere, removal, incoming, outgoing, etc.* used literally? The examples Lakoff and Johnson (1980, 1999) cite are more obviously metaphorical: they found that conceptual domains like time, the mind or emotions are often understood in terms of very different domains such as space (*Time Is Space*), substances (*Mind Is a Machine*) or forces (*Love Is a Physical Force*). In the cognitive linguistics literature, conceptual metaphors like these are referred to as *ontological metaphors*. In these conceptual metaphors phenomena are conceived in terms of ontologically different types of phenomena. The conceptual metaphor *Atmosphere Is Container* of course does not change the ontological domain: The atmosphere and a container are both construed in a spatial domain. Elements of the container schema are: an inside, an outside and a bounding surface. Rooms and houses are obvious containers: The walls, ground and roof are the boundaries; through the doors, we can move from the inside to the outside of the container etc. But even where there is no natural physical boundary that can be viewed as defining a container, we can conceptually impose boundaries: A national territory has an inside, a borderline and neighbours outside the country. The same holds for the *atmosphere*. The atmosphere does not have discrete boundaries; it has no top (i.e. it is just a model), no sides and no bottom (i.e. the gaseous atmosphere reaches deep into the ground).

Imagination typically requires us to impose artificial boundaries that make physical phenomena discrete—just as we are, entities bounded by a surface (Lakoff & Johnson, 1980, p. 30). We use this imaginative thinking when we construct our understanding of the atmosphere based on the container schema. The atmosphere is thought of as having a top made of *ozone* or $CO_2$ and conceived with energy flows in and out of this container. The same holds for conceptual metaphors like *Neuron Is Container*, for example. A neuron is a cell and, therefore, conceptualised within a spatial domain. But as a neuron is an object of the microcosm it is not open for direct experience. Therefore, it is in line with conceptual metaphor theory and its epistemological foundations that we understand the spatial structure of a nerve cell metaphorically: Even the term 'cell' itself is metaphorical as it refers to a monk's cell in a monastery. What happens here is that a conceptual metaphor is constructed by drawing on mesocosmic experience conceptualised in terms of image schematic structures to understand spatial structures in micro- (neuron) and macrocosm (atmosphere).

By mapping all aspects of these image schemata, transfers may occur that hinder an adequate conceptual understanding. Students often compare phenomena within the same ontological domain in terms of surface similarity rather than in terms of deeper relational structure (Holyoak & Koh, 1987). This supports the finding of Halpern, Hansen, and Riefer (1990) who found that near domain analogies cause more obstacles to understand a scientific concept than distant domain analogies. When the similarity between two phenomena is more obscure students are required to put more effort into mapping the underlying relationships in order to render it meaningful.

We see the extensive use of image schemata, even from the same ontological domain as the target to be construed, as support for the hypotheses that understanding needs to be grounded in mesocosmic experience. To reveal the underlying image schemata for understanding micro- and macrocosmic phenomena, conceptual metaphors have worked as a fruitful grain size in our study. Relating the conceptual metaphors students and scientists construct to understand phenomena reveals the students' learning demand.

*External Representations Informed by Conceptual Metaphor Theory*

Based on the conceptual metaphors presented in Table 5, we developed external representations to address students' alternative conceptions. To do so, we formulated the learning demand based on the gap between the conceptual metaphors of students and those of scientists (Table 7).

The students' learning demand analysed in our study can be separated into two different types of requirements. First, some alternative conceptions occur as a result of students' repeated experiences with phenomena of their everyday world and an inadequate mapping of an image schematic structure. Second, other alternative conceptions can be traced back to missing experiences, which have to be made during

Table 7.   Addressing students' experiential demand via external representations[a]

| Topic | Learning demand | External representations |
|---|---|---|
| Microbial growth | Understand that cell division consists of division and enlargement: Reflect on how division schema is employed | External representation 'Tearing paper': Divide a sheet of paper as a representation of the division schema |
| Saltatory signal conduction | Understand that myelin makes the action potential jump from node to node: Reflect on how start-path-goal schema is employed | External representation 'Toppling Dominos': Domino-brick and straw model |
| Greenhouse effect | Understand the role of $CO_2$ in climate change: Experience the properties of $CO_2$ and reflect on how container schema is employed | External representation 'Greenhouse effect' to afford experience on the role of $CO_2$ in global warming, reflect on the absence of ozone |
| | Understand the energy flows in global warming: Reflect on how balance schema is employed | External representation 'Visualise balance schema' to disclose and work with an implementation of the combined container and balance schemata, reflect its mapping to the dynamic equilibrium within the greenhouse effect |
| Carbon cycle | Understand that a constant $CO_2$ level means a balance in emission and removal: Reflect on how balance schema is employed | External representation 'Visualise balance schema' to disclose and work with an implementation of the combined container and balance schemata, reflect its mapping to the dynamic equilibrium within the carbon cycle |

[a]A full list of the external representations we analysed in our study is presented in Appendix.

science teaching. In the conceptual change framework these two approaches are discussed as 'misconceptions' and 'missing conceptions' (Aufschnaiter & Rogge, 2010).

With these requirements in mind the external representations presented in our study can be separated into two categories:

- External representations that address the experiential demand:
The conceptual metaphors students used to construe the greenhouse effect showed that they lack an adequate idea of the role of $CO_2$ in global warming. In this case, no or inadequate conceptions can be traced back to a missing experience of the phenomenon; actually the learning demand reveals an experiential demand. To deal with this we provided a mesocosmic experience (simulate the greenhouse effect) to present the properties of $CO_2$.

There are multiple representations that afford experiences of second-hand origin, such as photomicrographs, electromicrographs, chromatograms, recordings of action potentials and a view of a DNA sequencing gel. These representations, whether of first- or second-hand origin, can prepare the ground for the development of conceptions. Empirical methods in science are often means for students to experience beforehand imperceptible entities with the help of technical devices, for example, a microscope or a chromatograph. Representations that afford an experience of a phenomenon to be scientifically understood are of great importance for students. With an eye on the importance of experience, Johnstone (2007) demanded that every science lesson should start with the use of tangible experiences only. However, review studies indicate that making and interpreting scientific experiences in classrooms is a challenging task for students (Hofstein & Lunetta, 1982; Tobin, 1990). It seems that providing experiences to students does not always produce the intended motivation and understanding. In our approach, the analysis of students' conceptual metaphors was a prerequisite for the design of external representations that afford the essential experience.

- external representations that disclose the image schematic structure of concepts: In the cases of understanding microbial growth, saltatory signal conduction, the atmospheric energy budget and the $CO_2$ budget the students' conceptual metaphors reveal that they refer to the same image schemata as scientists. Divergences in the conceptions are due to a difference in mapping this image schematic structure to the target domains. Tearing paper, working with and reflecting on toppling dominoes and water flowing through a beaker are material representations of image schemata that students and scientists employ in understanding cell division, saltatory signal conduction, the carbon cycle or the greenhouse effect. These material representations of cognitive schemata helped students to re-experience the inherent structure of the schema, identify its essential elements and reflect on how they employ it in their effort to understand the phenomenon. This category of representation sheds light on the embodied conceptions that shape students' conceptual understanding. The external representations we developed realise the proposal of Amin (2009) that conceptual metaphor theory can inform the identification of a concept's image schematic grounding and reflecting on it. Models in classrooms often work in such a way that they provide new experiences students may use as a source for understanding. Representations that visualise an image schema and its mapping on a scientific concept work differently. They do not provide new experience; they induce an instance of a relived embodied experience. By working with these external representations students have the chance to analyse the structure of this specific experience and reflect on their embodied cognition.

In which category an external representation falls depends on how it is implemented in science teaching. The example of simulating the greenhouse effect shows how a single setting (affording experience on the properties of $CO_2$ in a glass container/ beaker) can, on the one hand, address a student's experiential demand and, on the other hand, helps him to reflect on the usage of an image schema. Therefore, the instructions given when working with the external representations are crucial. With a focus on addressing the experiential demand, tasks to observe and explain are helpful, while focussing on mapping often requires instructions to compare, to map or to analogise. For the latter case, the example of Lena on the carbon cycle is typical. Often students explicitly need to be asked to map their experience to the phenomenon to be understood. Therefore, not only the external representation itself is crucial, but also how it is implemented plays a major role for it to be fruitful.

*Raising Metaconceptual Awareness by Reflecting on Image Schemata*

When analysing the students' performance during the teaching experiments, our attention was drawn to the fact that after working with the external representations on the greenhouse effect and signal conduction, some students related the newly constructed conceptions to everyday life contexts—without being prompted to do so. In the case of signal conduction, a student saw saltatory signal conduction as passing the ball while playing handball; in the case of understanding the greenhouse effect, a student related the in- and outflow of energy in the atmosphere to his personal budget. These kinds of student-generated mappings are discussed in science education literature as self-generated analogies (Aubusson & Fogwill, 2006) or spontaneous analogies (Haglund & Jeppsson, 2012).

To us, the case where a student construes his weekly budget based on the same image schema of a balance like the atmospheric energy system is especially interesting as several authors report evidence that an adequate understanding of stock-and-flow relationships in science or everyday life is very rare (Cronin, Gonzalez, & Sterman, 2009; Sweeney & Sterman, 2000). As this analogical mapping of an image schema to both science and everyday life contexts was only an incidental finding, we are far from a sound generalisation of this finding, but we interpret it as an indicator of students' metaconceptual thinking. Mason (1994) has pointed out that successful analogical reasoning depends to a great extent on the metacognitive awareness of the nature and purpose of the mapping. She defines metacognitive competence as reflecting on what one knows and how new knowledge is developed by integrating it with the pre-existing conceptions.

We found other situations where working with the external representations that were developed based on the students' conceptual metaphors raised their awareness of their own conceptual status and progress: e.g. 'I thought that the ozone hole is responsible for warming. But it cannot be ... ' (Ann on the greenhouse effect); 'The idea "Warming By More Input" was what we initially thought ... ' (Max on the greenhouse effect). These different cases of metacognition can be interpreted by the kind of analysis used by Gilbert (2005) and Von Wright (1992), who discern two levels of metacognition in working with visualisations: At the lower level, an individual is capable of considering and comparing her conceptions to familiar contexts, whilst at the upper level she can reflect on her own knowledge. Adapted to our example of working with external representations, drawing analogical mappings to everyday life contexts is located on the lower level, while reflecting on the conceptual status (like the cases of Ann and Max above) indicates the higher level.

We draw back students' metaconceptual awareness to the type of external representations that encouraged them to reflect on their mapping of the image schemata to understand the incoming and outgoing radiation in the atmosphere or the processes in signal conduction. The external representation 'visualised balance schema' scrutinises the image schematic structure of understanding the energy flows in making the container image schema (beaker) and the balance schema (relation of inflow/ outflow) explicit. This external representation is not only a representation of the phenomena of climate change but it is also an external representation of the container and the balance schema. The reflection on the structure of these image schemata seemed to support the students in understanding the atmospheric energy balance (often referred to as an energy *budget*), on the one hand. On the other hand, it seemed to make them aware of stock-and flow-relationships in their everyday life, too. Gilbert (2005) pointed out that becoming metacognitive is an important challenge to successfully deal with external representations like visualisations. The findings discussed above indicate that external representations that are designed to reflect on a concept's image schematic grounding supports students' metacognitive abilities.

## Conclusions

Many decades ago, a tradition of research emerged that collected students' conceptions to *describe* how students understand certain science concepts. In recent years, several researchers in science education adapted the theoretical framework of embodied cognition to science education to *explain* why students think the way they think, i.e. to understand students' understanding; experience is the pivotal process for the development of understanding. This experience takes place in the world of medium dimensions, which Vollmer (1984) calls the mesocosm. We adapted the theoretical framework of conceptual metaphors along with Vollmer's epistemological distinction of micro-, meso- and macrocosm to science education and found that these frameworks can serve as *diagnostic tools to predict* the degrees of students' difficulties in understanding. Because understanding is firmly grounded in experience and, thus, in the mesocosm, understanding needs to be rooted in mesocosmic experience. We took this central claim of conceptual metaphor theory to elaborate the *prescriptive value* of this theoretical framework. We hope to have shown via evidence, and argued via theory, how external representations that help students reflect on their embodied conceptions from the mesocosm can improve the understanding of science. In our teaching experiments, we found the notion of conceptual metaphors to be useful for science education in two ways: it can serve as a theory to analyse conceptions and it is helpful for the design of external representations. Or, thinking metaphorically, uncovering how the hidden hand of our mesocosmic, embodied conceptions guide our understanding sheds light on the nature of understanding. In this way, offering this hidden hand to science educators enables them to use it as a guiding hand to enable a deeper understanding.

## Acknowledgements

We thank Reinders Duit, David Treagust and the editors of this special issue for valuable and critical comments on an earlier draft of this paper. Their comments and suggestions enabled us to improve the quality of this contribution. Any remaining inadequacies are ours.

## Disclosure statement

No potential conflict of interest was reported by the author(s).

## Note

1. Note that the scale Johnstone proposed is different from the one Vollmer proposed (cf. Table 1). Johnstone calls the perceptible level *macroscopic*, while Vollmer points to the perceptible world as the *mesocosm*.

## References

Amin, T. G. (2009). Conceptual metaphor meets conceptual change. *Human Development, 52*(3), 165–197.

Amin, T. G., Jeppsson, F., Haglund, J., & Strömdahl, H. (2012). Arrow of time: Metaphorical construals of entropy and the second law of thermodynamics. *Science Education, 96*(5), 818–848.

Aubusson, P. J., & Fogwill, S. (2006). Role play as analogical modelling in science. In P. J. Aubusson, A. G. Harrison, & S. M. Ritchie (Eds.), *Metaphor and analogy in science education* (Vol. 30, pp. 93–104). Berlin: Springer.

Aufschnaiter, C. V., & Rogge, C. (2010). Misconceptions or missing conceptions? *Eurasia Journal of Mathematics, Science and Technology Education, 6*(1), 1–16.

Campbell, N. A., Reece, J., Urry, L., Cain, M., Wasserman, S., Minorsky, P., & Jackson, R. (2008). *Biology.* (8 th ed.). New York: Pearson Benjamin Cummings.

Cronin, M., Gonzalez, C., & Sterman, J. (2009). Why don't well-educated adults understand accumulation? A challenge to researchers, educators, and citizens. *Organizational Behavior and Human Decision Processes, 108*(1), 116–130.

Davidowitz, B., & Chittleborough, G. (2009). Linking the macroscopic and sub-microscopic levels: Diagrams. In J. K. Gilbert & D. F. Treagust (Eds.), *Multiple representations in chemical education* (Vol. 4, pp. 169–191). Dordrecht: Springer.

Diesterweg, A. (1835). *Wegweiser zur Bildung für deutsche Lehrer* [A Guide for German Teachers]. Essen: Baedeker Verlag.

Duit, R. (2009). *Bibliography—STCSE students' and teachers' conceptions and science education.* Retrieved from http://archiv.ipn.uni-kiel.de/stcse/

Duit, R., Gropengiesser, H., Kattmann, U., & Komorek, M. (2012). The model of educational reconstruction—a framework for improving teaching and learning science. In D. Jorde & J. Dillon (Eds.), *Science education research and practice in Europe* (pp. 13–38). Rotterdam: Sense

Duit, R., & Treagust, D. F. (2003). Conceptual change: A powerful framework for improving science teaching and learning. *International Journal of Science Education, 25*(6), 671–688.

Fauconnier, G., & Turner, M. (2002). *The way we think.* New York, NY: Basic Books.

Felzmann, D. (2014). Using metaphorical models for describing glaciers. *International Journal of Science Education, 32*(16), 2795–2824.

Fichtner, J. (2013). *Evaluation einer Unterrichtseinheit zur Neurobiologie am Beispiel Multiple Sklerose* [Evaluation of a teaching sequence on teaching neurobiology]. Lueneburg: Leuphana University.

Fuchs, H. U. (2007). *From image schemas to dynamical models in fluids, electricity, heat, and motion.* Zentrum Für Mathematik Und Physik, ZHaW.

Gilbert, J. K. (2005). Visualization: A metacognitive skill in science and science education. In J. Gilbert (Ed.), *Visualization in science education* (pp. 9–27). Dordrecht: Springer.

Gilbert, J. K., & Treagust, D. F. (2009). *Multiple representations in chemical education* (Vol. 4). Dordrecht: Springer. doi:10.1007/978-1-4020-8872-8

Gropengiesser, H. (1997). *Didaktische Rekonstruktion des 'Sehens'* [Educational Reconstruction of 'Seeing']. Oldenburg: ZpB Zentrum für pädagogische Berufspraxis.

Haglund, J., & Jeppsson, F. (2012). Using self-generated analogies in teaching of thermodynamics. *Journal of Research in Science Teaching*, 49(7), 898–921.

Halpern, D., Hansen, C., & Riefer, D. (1990). Analogies as an aid to understanding and memory. *Journal of Educational Psychology*, 82, 298–305.

Harrison, A., & de Jong, O. (2005). Exploring the use of multiple analogical models when teaching and learning chemical equilibrium. *Journal of Research in Science Teaching*, 42(10), 1135–1159.

Harrison, A., & Treagust, D. F. (2006). Metaphor and analogy in science education. In P. Aubusson, A. Harrison, & D. Ritchie (Eds.), *Metaphor and analogy in science education* (pp. 11–24). Dordrecht: Springer.

Hodson, D. (1990). A critical look at practical work in school science. *School Science Review*, 71(256), 33–40.

Hofstein, A., & Lunetta, V. N. (1982). The role of the laboratory in science teaching: Neglected aspects of research. *Review of Educational Research*, 52(2), 201–217.

Holyoak, K. J., & Koh, K. (1987). Surface and structural similarity in analogical transfer. *Memory & Cognition*, 15(4), 332–340.

IPCC. (2013). *Climate change 2013 – the physical science basis* (pp. 1–1552). Cambridge: Cambridge University Press.

Jeppsson, F., Haglund, J., Amin, T. G., & Strömdahl, H. (2013). Exploring the use of conceptual metaphors in solving problems on entropy. *Journal of the Learning Sciences*, 22(1), 70–120.

Johnson, M. (1987). *The body in the mind—the bodily basis of meaning, imagination, and reason.* Chicago, IL: University of Chicago Press.

Johnstone, A. H. (1982). Macro and microchemistry. *School Science Review*, 19(3), 71–73.

Johnstone, A. H. (2007). *Science education: We know the answers, let's look at the problems.* 5th Greek Conference Science Education and new technologies in education, Athens, pp. 1–13.

Kattmann, U., Duit, R., & Gropengiesser, H. (1998). The model of educational reconstruction—bringing together issues of scientific clarification and students' conceptions. In H. Bayrhuber & F. Brinkmann (Eds.), *What–why–how? Research in Didaktik of biology. Proceedings of the first conference of European researchers in Didaktik of biology (ERIDOB)* (pp. 253–262). Kiel: IPN.

Komorek, M., & Duit, R. (2004). The teaching experiment as a powerful method to develop and evaluate teaching and learning sequences in the domain of non-linear systems. *International Journal of Science Education*, 26(5), 619–633.

Lakoff, G., & Johnson, M. (1980). *Metaphors we live by.* London: The University of Chicago Press.

Lakoff, G. (1990). *Women, fire, and dangerous things: What categories reveal about the mind.* Chicago, IL: University of Chicago Press.

Lakoff, G., & Johnson, M. (1999). *Philosophy in the flesh.* New York: Basic Books.

Lakoff, G., & Núñez, R. (2000). *Where mathematics comes from.* New York, NY: Basic Books.

Mathewson, J. (2005). The visual core of science: Definition and applications to education. *International Journal of Science Education*, 27(5), 529–548. doi:10.1080/09500690500060417

Mason, L. (1994). Analogy, metaconceptual awareness and conceptual change: A classroom study. *Educational Studies*, 20(2), 267–291.

Mayring, P. (2002). Qualitative content analysis—research instrument or mode of interpretation? In M. Kiegelmann (Ed.), *The role of the researcher in qualitative psychology* (pp. 139–148). Tübingen: UTB.

Nelson, P. G. (2002). Teaching chemistry progressively: From substances, to atoms and molecules, to electrons and nuclei. *Chemistry Education Research and Practise*, 3(2), 215–228. doi:10.1039/B2RP90017C

Niebert, K. (2007). Wachsen Haare schneller, wenn man sie öfter schneidet? In H. Vogt, D. Krueger, & S. Marsch (Eds.), *Erkenntnisweg Biologiedidaktik* (pp. 7–22). Rostock: Universität Rostock.

Niebert, K., & Gropengiesser, H. (2013a). The model of educational reconstruction: A framework for the design of theory-based content specific interventions. The example of climate change. In T. Plomp & N. Nieveen (Eds.), *Educational design research—Part B: Illustrative cases* (pp. 511–531). Enschede: SLO (Netherlands institute for curriculum development).

Niebert, K., & Gropengiesser, H. (2013b). Understanding and communicating climate change in metaphors. *Environmental Education Research, 19*(3), 282–302.

Niebert, K., & Gropengiesser, H. (2014). Understanding the greenhouse effect by embodiment–analysing and using students' and scientists' conceptual resources. *International Journal of Science Education, 36*(2), 277–303.

Niebert, K., Marsch, S., & Treagust, D. F. (2012). Understanding needs embodiment: A theory-guided reanalysis of the role of metaphors and analogies in understanding science. *Science Education, 96*(5), 849–877.

Núñez, R. E., Edwards, L. D., & Filipe Matos, J. (1999). Embodied cognition as grounding for situatedness and context in mathematics education. *Educational Studies in Mathematics, 39*(1), 45–65.

Riemeier, T., & Gropengiesser, H. (2008). On the roots of difficulties in learning about cell division: Process-based analysis of students' conceptual development in teaching experiments. *International Journal of Science Education, 30*(7), 923–939.

Rohrer, T. (2001, July). *Understanding through the body: fMRI and ERP investigations into the neurophysiology of cognitive semantics*. Presented at the 7th International Cognitive Linguistics Association, University of California at Santa Barbara, CA.

Rohrer, T. (2005). Image schemata in the brain. In B. Hampe & J. Grady (Eds.), *From perception to meaning: images schemas in cognitive linguistics* (pp. 165–196). Berlin: Mouton de Gruyter.

Schmitt, R. (2005). Systematic metaphor analysis as a method of qualitative research. *The Qualitative Report, 10*(2), 358–394.

Schneeweiss, H. (2008). *Biologie verstehen: Bakterien*. [Understanding Biology: Bacteria]. Oldenburg: Didaktisches Zentrum.

Semino, E. (2008). *Metaphor in discourse*. Cambridge: Cambridge University Press.

Steffe, L., & Thompson, P. (2000). Teaching experiment methodology: Underlying principles and essential elements. In A. Kelly & R. Lesh (Eds.), *Handbook of research design in mathematics and science education* (pp. 277–309). London: Lawrence Erlbaum Associates.

Sterman, J. D. (2008). Risk communication on climate: Mental models and mass balance. *Science, 322*(5901), 532–533.

Sweeney, L. B., & Sterman, J. D. (2000). Bathtub dynamics: Initial results of a systems thinking inventory. *System Dynamics Review, 16*(4), 249–286.

Tobin, K. (1990). Research on science laboratory activities: In pursuit of better questions and answers to improve learning. *School Science and Mathematics, 90*(5), 403–418.

Tsui, C.-Y., & Treagust, D. F. (2013). Introduction to multiple representations: Their importance in biology and biological education. In D. F. Treagust & C.-Y. Tsui (Eds.), *Multiple representations in biological education* (pp. 3–18). Dordrecht: Springer Netherlands.

Van Someren, M. W., Reimann, P., Boshuizen, H. P. A., & de Jong, T. (1998). *Learning with multiple representations*. London: Elsevier.

Vollmer, G. (1984). Mesocosm and objective knowledge. In F. Wuketits (Ed.), *Concepts and approaches in evolutionary epistemology* (pp. 69–121). Dordrecht: Reidel Publishing Company.

Von Wright, J. (1992). Reflections on reflection. *Learning and Instruction, 2*(1), 59–68.

Wilson, M. (2002). Six views of embodied cognition. *Psychometric Bulletin & Review, 9*(4), 625–636.

**Appendix. From Embodied Conceptions to External Representations**

Full list of students' and scientists' embodied conceptions, the resulting learning demand, and the external representations we developed based on conceptual metaphor theory in our study.

| Topic | Students' conceptual metaphors | Scientists' conceptual metaphors | Learning demand | External representations |
|---|---|---|---|---|
| Cell Biology | Dividing Is Becoming More | Dividing Is Becoming More / Dividing Is Becoming Smaller | Understand that cell division consists of division and enlargement: Reflect the division schema | External representation 'Tearing paper': Divide a sheet of paper as a representation of the division schema |
| | Growth Is Division | Growth Is Division and Enlargement | | |
| | Growth Is Becoming Mature | Growth Is a Cell Division | Understand the cellular principles of growth | External representation 'Onion roots', external representation 'Microscope' Observe the growth of onion roots with a naked eye (mesocosm) and microscope the root cells (microcosm) |
| | Cell Is Flat Structure | Cell Is Bodily Structure | Understand that a cell is a structure in three dimensions instead of two | External representation 'Soap bubbles': Comparing the 2D/3D relations of viewing a cell under a microscope with observing the structure of soap-bubbles in an aquarium via a glass wall (2D) or from top (3D) |
| | Gene Is Containing Information | Gene Is Information | Understand the ontology of a gene | External representation 'DNA sequence': Original data sheets with DNA sequences |
| | DNA Is Containing Code, Code Is Sequence of Numbers | DNA Is Code, Code Is Sequence of Bases | Reflect the conception codes | |

(Continued)

**Appendix.** Continued

| Topic | Students' conceptual metaphors | Scientists' conceptual metaphors | Learning demand | External representations |
|---|---|---|---|---|
| | Organism Is Containing Cells | Organism Is Made of Cells | Reflect on the ontology of cells | External representation 'Microscopy': Microscopy of root cells; reflection of part–whole image schema |
| | Division Is Cutting Information | Division Is Doubling Information | Understand the replication of a genome during mitotic cell division | External representation 'Tearing manual': Compare the tearing of a construction manual with genome division |
| Neurobiology | Conduction Speed Is Depending on Range of Path | Conduction Speed Is Depending on Range of Path | Understand that myelin makes the action potential jump from node to node: Reflect the travel schema | External representation 'Toppling Dominos': Domino-brick and straw model |
| | Signal Is Travelling Agent | Conduction Speed Is Depending on Speed of Agent | Understand the isolating role of myelin | External representation 'Myelin': Electromicroscopic photos of myelinated and demyelinated neurons |
| | Neuron Is Container | Signal Is Jumping Agent  Neuron Is Container, Myelin Is Boundary of Container | | |
| Greenhouse effect | Greenhouse Effect Is More Input, Greenhouse Effect Is Less Output | Greenhouse Effect Is Shifted Equilibrium | Understand the energy flows in global warming: Reflect the balance schema | External representation 'Reflect balance schema' to disclose and work with an implementation of the combined container and balance schemata, reflect its mapping to the dynamic equilibrium within the greenhouse effect |
| | CO$_2$ Is Detrimental, CO$_2$ Captures Heat | CO$_2$ Is Capturing and Releasing Heat  Atmosphere Is Container | Understand the role of CO$_2$ in climate change: Experience the properties of CO$_2$ and reflect on the container schema | External representation 'Greenhouse effect' to afford experience on the role of CO$_2$ in global warming, and reflect on the absence of ozone |

| | | | | |
|---|---|---|---|---|
| | Atmosphere Is Container | Atmosphere Is Container $CO_2$ Is Content | Understand the role of $CO_2$ in climate change: Reflect on the container schema | External representation 'Greenhouse effect' Measure the temperature on the bottom, in the middle and on top of a beaker in the greenhouse experiment |
| | $CO_2$ Is Top of Container, $CO_2$ Is Destroyer of Boundary $CO_2$ Is Reflecting Heat | $CO_2$ Is Opaque for Heat $CO_2$ Is Transparent for Light | Understand that $CO_2$ reacts differently with light and heat | External representation '(Im)permeable $CO_2$': Two plastic bags, one filled with air and the other filled with $CO_2$, are illuminated with a light bulb on one side. The brightness and temperature are measured on the other side |
| Carbon Cycle | Carbon Pools Are Containers | Carbon Pools Are Containers | Understand the nature of carbon containers | External representation 'Carbon pools' Present carbon containing materials (plants, air, sea water, molluscs, oil, etc.) |
| | $CO_2$ Is Man-Made, $CO_2$ is Man-Made or Natural | Carbon Flow Is Manmade or Natural | Understand $CO_2$ as a natural element of the atmosphere | External representation 'Track record' Historical track record with $CO_2$ in the atmosphere |
| | Climate Change by man-made $CO_2$ | Climate Change By Imbalance in the carbon cycle | Relate climate change to manmade carbon flows instead of manmade $CO_2$ particles | External representation '$CO_2$- Molecule': Molecular model of a $CO_2$- molecule and external representation 'Container-ball model' to model carbon flows |
| | Carbon Flows Are One Way Only | Carbon Flows Are Cyclic | Understand that in cyclic processes the start-path-goal schema is transferred into a cycle schema | External representation 'Container-ball model': Model carbon flows in a container-ball model to reflect on start-path-goal schema and cycle schema |
| | Constant $CO_2$ level Is Constant Input, Constant $CO_2$ level Is Constant Output | Carbon Pools Are Containers Constant $CO_2$ level By Balanced Input and Output | Understand that a constant $CO_2$ level means a balance in emission and removal: Reflect balance schema | External representation 'Reflect balance schema' to disclose and work with an implementation of the combined container- and balance schemata, reflect its mapping to the dynamic equilibrium within the cycle |
| | Carbon Flows Are Balanced | Carbon Flows Are Series of Imbalances | Understand that in the carbon cycle a series of imbalances creates a balanced carbon budget | External representation 'Container-ball model with multiple flows': Reflect the mapping of the balance schema in a set of multiple flows |

# From Stories to Scientific Models and Back: Narrative framing in modern macroscopic physics

Hans U. Fuchs

*Institute of Applied Mathematics and Physics, School of Engineering, Zurich University of Applied Sciences at Winterthur, Winterthur, Switzerland*

Narrative in science learning has become an important field of inquiry. Most applications of narrative are extrinsic to science—such as when they are used for creating affect and context. Where they are intrinsic, they are often limited to special cases and uses. To extend the reach of narrative in science, a hypothesis of narrative framing of natural and technical scenes is formulated. The term narrative framing is used in a double sense, to represent (1) the enlisting of narrative intelligence in the perception of phenomena and (2) the telling of stories that contain conceptual elements used in the creation of scientific models of these phenomena. The concrete case for narrative framing is made by conceptual analyses of simple stories of natural phenomena and of products related to modern continuum thermodynamics that reveal particular figurative structures. Importantly, there is evidence for a medium-scale perceptual gestalt called FORCE OF NATURE that is structured metaphorically and narratively. The resulting figurative conceptual structure gives rise to the notion of natural agents acting and suffering in storyworlds. In order to show that formal scientific models are deeply related to these storyworlds, a link between using (i.e. simulating) models and storytelling is employed. This link has recently been postulated in studies of narrative in computational science and economics.

## Introduction

The investigation described here grew out of a number of questions all centrally related to how humans understand nature and (natural) science. Along the way, a unified approach to the physics of dynamical systems and a theory of uniform thermal dynamical processes—all based upon modern continuum physics—were developed (Dumont, Fuchs, Maurer, & Venturini, 2014; Fuchs, 2010); cognitive linguistics has been employed to study the conceptual structures embedded in macroscopic physical science; stories of forces of nature have been produced and used for application in teacher training and in a novel primary school curriculum in Italy; and lately, the question has been taken up of how small-scale embodied conceptual structures (such as conceptual metaphor) relate to the large-scale structures of story and storyworld.

In the course of these studies, a number of points have become clear. Different fields of continuum physics[1] and the physics of macroscopic physical science all make use of the same few basic figurative structures allowing us to write theories and models in strongly analogous forms. It is possible to summarize these structures as a network of (small-scale) figures of mind[2] that lead to the conceptualization of a (medium-scale) perceptual gestalt I call FORCE OF NATURE.[3] The perception of concrete forces of nature (wind, water, light, ice and fire, electricity, motion, food, or soil, to name but a few) leads us to construct this conceptual network as a matter of everyday life—the figures of mind used to understand folk physics are also those that structure modern macroscopic physical science.

These observations lead me to propose a hypothesis regarding the relation between narrative and science. The perception of forces of nature in large-scale events (a winter storm) lets us construct the figure of *natural agents* central to stories (this is a process of *narrative perception*). This, in turn, allows us to tell stories about forces of nature (as an act of *narrative production*). On the other hand, recipients of such stories can build storyworlds having a certain form in which natural agents act and suffer according to certain rules (this is again an act of *narrative perception*).

So far, this should not come as a surprise—the particular way it is phrased may sound novel but it is simply a description of humans interacting with the world and creating semiotic products of folk science. My question is how this relates to formal science. Remember that the cognitive model of FORCE OF NATURE can be shown to be fundamentally the same in formal macroscopic physics and in folk science. Therefore, we can conjecture that narrative perception and narrative production also apply to the products of formal science—if only indirectly. We will see in this paper that the relation between the simulation of formal models and the act of storytelling will help us create a link between the roles of narrative in everyday understanding of the natural world (folk science) and science.

In this paper, I will refer to the various processes and acts of narrative perception and narrative production as the *narrative framing* of natural scenes.[4] We can understand this term as referring to two different but intimately related senses of the use of narrative: (1) enlisting of narrative intelligence in the perception of phenomena[5] (we perceive living through a winter storm as a story, and when we hear stories or are exposed to formal models and their simulations, we also bring narrative perception to bear upon the understanding of the semiotic products) and (2) producing and telling stories that contain conceptual elements used in creating formal models of these phenomena (we narrate stories about the winter storm using language that contains the seeds of the concepts of which formal scientific models are made). The former is equivalent to saying that we have a narrative mind, a point that can only be inferred indirectly by evidence gained from our behavior as observers of nature and recipients and creators of stories. The latter is accessible to direct observation: we can study semiotic products used both in everyday life and in science and make the case for the narrative nature of our concepts in folk science and in formal science (and show that they are closely related).

This paper is structured as follows. In the next section, some background material will be outlined (mainly relating to issues such as conceptual metaphor and narrative). Then, in order to introduce the reader to the range of semiotic phenomena available to us, three examples will be presented—a Winter Story for small children, Sadi Carnot's narrative introducing and motivating his model of the power of heat, and a formal model of electrical heating of water in a teakettle. A reading of these examples reveals conceptual metaphoric and narrative structures, most importantly those used to give form to the *gestalt of force of nature*.

These first parts of the paper prepare the ground for a more general discussion of narrative framing in the light of narratology (as cognitive science) and recent studies of narrative in science—computational science and economics, to be precise. We will find it useful to compare the relationship between simulations and models to that between stories and *storyworlds*. There we will see how models (a major component of scientific work) relate to storyworlds (the cognitive models created by the perception of narratives).

Research into the subject dealt with in this paper has greatly profited from cognitive science in general and cognitive linguistics in particular. Therefore, in the Conclusion, I will return to the wider concern of research in embodied cognition in science and discuss how the present paper may add to this endeavor. Moreover, reference to work in the field of narrative in science education will be made and we will see how the idea of narrative framing relates to studies of the use of narrative in the science classroom. Finally, some applications of the extended use of narrative being conducted at present will be mentioned as a view to the future of research in this field.

## Metaphors and Narratives

In order to prepare the reader for some of the terminology used, we will first take a brief look at conceptual metaphor theory and the postulate of embodied cognition, and a modern theory of narrative.

### Embodied Cognition and Conceptual Metaphor

The tools used in the analyses presented in this paper have been developed in cognitive science, particularly in cognitive linguistics and in work revolving around the model of an embodied mind. A leading model in the science of the human mind, the one I adopt here, has evolved from an integral view of the interaction of human organisms with their natural, social, and psychological environments. The embodied mind is assumed to be a product of this interaction (Chemero, 2009; Dewey, 1925; Gibbs, 2006; Gibson, 1966, 1979; James, 1890/1983; Johnson, 1987; Lakoff and Johnson, 1999; Noë, 2004). Simply put, in a dynamical systems view of embodied cognition (Thelen and Smith, 1994), the nervous system of the organism resonates with its body in its interactions with the environment, leading to dynamical patterns which can variously be described as basins of attraction or, more usefully for us, shapes or gestalts (Arnheim, 1969; Johnson, 1987). As constructs of perception, such shapes are projected to lead to new structures, that is, figures of mind such as conceptual metaphor (see the later text). Our representations of the outside world are not direct reflections but rather representations of the figures given to us by perception and imagination.

The assumption of an *embodied mind* has important consequences for the form and meaning of our linguistic creations. What is often seen as literal language is, on scrutiny, often found to be implicitly figurative—mirroring a figurative mind. Cognitive linguistics has become a tool for investigating the human mind as it is reflected in our language. For example, metaphoric expressions are no longer viewed as embellishments of language but rather as expressing deep-rooted forms of understanding of the world (Gibbs, 1994; Johnson, 1987; Lakoff, 1987; Lakoff and Johnson, 1980, 1999; Lakoff & Núñez, 2000; Talmy, 2000a, 2000b).

*Cognitive linguistics* and the model of an embodied mind have also been the starting point for identifying image schemas as gestalts abstracted from recurring experience of bodily interactions with the environment (Hampe, 2005; Johnson, 1987; Lakoff, 1987; and see Arnheim, 1969, whose discussion of the 'intelligence' of visual perception leads to a similar point). Among image schemas, we find CONTAINER, PATH, (FLUID) SUBSTANCE, SCALE, BALANCE, PROCESS, and the schemas identified in force dynamics and spatial relations by Talmy (2000a) and Langacker (1987, 1991). Many of the most basic forms of conceptualizations used and reflected in language are based upon metaphoric projections of these types of schemas, such as when we say that 'heat has been collecting in the room' (note the schemas of CONTAINER and FLUID SUBSTANCE in this example).

*Conceptual metaphor* theory, as a branch of cognitive linguistics, makes an important distinction between CONCEPTUAL METAPHOR and metaphoric linguistic expression. The latter is what we hear or read when somebody uses a metaphor, the former is a figure of mind—we might say it is the actual concept. For example, 'heat flows through the walls of the building' is an example of an expression for the underlying metaphor HEAT IS A FLUID SUBSTANCE. Metaphors are the result of the (metaphoric) projection (Turner, 1996) of structure from a source domain onto conceptual structure in a target domain. A primary form of metaphor can result if structure from an image schema is projected—such as in the examples used in the previous paragraph. Primary metaphors can be combined into complex metaphors by conceptual blending (Lakoff and Johnson, 1999).

## Theory of Narrative: Stories and Storyworlds

In this section, I provide a brief description of a couple of elements from recent research in narratology. The points raised are somewhat different from discussions of narrative and storytelling in science and science education (see Norris, Guilbert, Smith, Hakimelahi, & Phillips, 2005, for an important contribution to the latter). The goal is to understand enough of the theory of narrative for us to discuss the role of 'good' stories in the model of narrative framing.

*A model of narrative.* In this paper, I will use a model of narrative as a *radial category* (Herman, 2009). It allows us to consider different forms of narrative as belonging to the same category; at the same time it tells us more clearly what we mean by *story*. According to a modern theory of categorization (Lakoff, 1987; Rosch, 1973), a radial category is one that has central or prototypical members (the categories exhibit prototype effects) and members that do not share that status—they are less prototypical and more peripheral. An example is the category of chair, with a typical dining room chair as a central member and a beanbag chair as a rather untypical one.

Herman calls *story* the prototypical member of the *category of narrative*. There are non-central members that relate to the categories of *description* and *explanation*, so-called narrativized descriptions and descriptivized narrations, or explanatory narratives and narrativized explanations. Briefly, narrativized descriptions and descriptivized narrations are (text) types between description and (prototypical) narrative; explanatory narratives and narrativized explanations are (text) types between prototypical narrative and explanation with varying degrees of emphasis on either narrative or explanation (see Herman, 2009, pp. 89–100, for more details). According to Herman (2009), four elements constitute the central member of the category of narrative. I will recount his list in slightly different words. Stories are narratives that include *all of the following elements*: (1) *events*; (2) (conscious) *experiencing of events* by *agents*; (3) *tension for creating events*; and (4) reason or *occasion for telling* by a narrator. Herman argues strongly for the roles of intentionality and author in stories (Herman, 2013).

*Stories and storyworlds.* The distinction between stories and storyworlds will prove important in the following. Stories are concrete narratives, whereas storyworlds are the mental models we construct when we are exposed to stories—stories transport us into storyworlds.

In *Story logic,* Herman (2002) defines storyworlds as follows: '[ ... ] storyworlds [are construed] as mental models [ ... ] supporting narrative understanding.' (p. 17). He writes that:

> [i]n trying to make sense of a narrative, interpreters attempt to reconstruct not just what happened—who did what to or with whom, for how long, how often and in what order—but also the surrounding context or environment embedding existents, their attributes, and the actions and events in which they are more or less centrally involved. [ ... ] storyworld points to a way interpreters of narrative reconstruct a sequence of states, events, and actions not just additively or incrementally but integratively or 'ecologically' ... (pp. 13–14)

To put this more simply and directly for our purpose, a story recounts the *what* of events and the storyworld we construct informs us about the *why*. Applying this distinction to the relation between narrative and science, we will be able to refer to stories as simulations and storyworlds as the models simulated (see the later text).

## A Winter Story and Carnot's Power of Heat

Can scientific narratives be proper stories? Can such stories help us understand nature and build scientific models? Can we use narrative intelligence to understand scientific models? To discuss these and related questions, I will describe three examples constructed for widely different applications—a story for primary science for small children and a word model of the operation of heat in steam engines by Sadi Carnot (this section), and a dynamical model of electric heating of water in a teakettle that uses the uniform dynamical models version of continuum thermodynamics (the following section).

### *A Winter Story*

Here is a shortened version of a story that was originally written to investigate questions relating to narrative and science learning (Fuchs, 2011, 2013a) and has since been used in the training of teachers and in early elementary school (Corni, 2013; Corni, Giliberti, & Fuchs, 2013). The story narrates how cold holds a wintry town in its grip.

> A small town called Little Hollow lay in a hollow surrounded by a high plain. As the last of the warmth of late fall left the plain surrounding Little Hollow, the cold of winter found its way into the area and spread out. So it was not all that cold up there. Even in the midst of winter, the sun managed to send some warming rays onto the plain. The snow that fell on the plain was not so cold either, but it was plenty, and the people of Little Hollow loved to go up to the plain for cross country skiing.
>
> But in Little Hollow, things were different. The cold of winter knew a good place where it could do its job of making everything and everybody cold much more easily. It could flow into the hollow where the town had been built. It could collect there and it knew it would not be driven out so easily by a little bit of wind as could happen on the plain. More and more cold could collect in Little Hollow, and it got colder and colder as the winter grew stronger. The temperature fell and fell.
>
> The people of Little Hollow knew that the cold would find its way into their homes if they were not careful to close windows and doors. The cold could even sneak in through tiny cracks between walls and windows, so the people had learned to build their homes well to make it hard for cold to flow in. At times when much cold had collected in their town, when it had become terribly cold and the temperature was very, very low, the fires in the furnaces had to work very hard to fight the cold. The people in their homes made sure that the heat produced by the furnaces would always balance the cold so that their homes felt comfortably warm.
>
> For the children of Little Hollow, the cold of winter was not so bad. They dressed warmly and played hard when they were outside. But even for them, the thick cold of winter had mischief in mind. It went into the snow lying on the ground to make it very cold as well and this made the snow drier and harder to work with. The children could not form

snowballs, and it was much more difficult to build snowmen. They had to wait until winter had grown somewhat tired, and the cold was slowly driven out of Little Hollow. When that happened the cold of winter knew its time had come. The warmth of early spring would grow stronger and drive the cold out of the hollow. The cold knew it had to accept its defeat but it also knew very well it would be back...

If a story should be prototypically narrative, it must make use of and be shaped by agents, tension, events and processes, causation/power, and connection to emotional understanding (Herman, 2009, 2013). If it is to contribute to scientific thinking, it must also contain the small-scale conceptual structuring of storyworld and agents that make scientific formal reasoning possible. Such structuring can be provided by metaphoric projection of schemas (Johnson, 1987; Turner, 1996). Our story appears to meet both criteria.

*Analysis of the Winter Story*

The Winter Story creates a storyworld: it describes a scene relative to which natural agents (cold and heat) are profiled (their characters are outlined). Moreover, by being embedded in a story, cold is given a character with emotional aspects[6] (related to the generating polarity, i.e. COLD ↔ HOT) causing and being subject to processes that unfold over time. The agents act and suffer in accordance with their properties (characters), and the story narrates a particular course of events in this storyworld.

The storyworld receives structure from the conceptualization of its elements—in particular, the agents appearing in it—in terms of figures of mind. We can identify a list of *image schematic elements* whose *metaphoric projection* leads to the fleshing out of a character or *agent* called *cold* (see a selection of expressions from the story in Table 1). Table 1 lists the types of conceptual metaphors that characterize forces (of nature): the THERMAL LANDSCAPE metaphor, the MOVING COLD metaphor, and the COLD AS A MOVING FORCE metaphor (I am using names analogous to Johnson's three metaphors in his analysis of music; Johnson, 2007, pp. 248–254). These metaphors reflect our figurative understanding of the major properties of forces (of nature). To be more specific, the story creates partial models of cold (such as when the cold outside tries to sneak into a home through the walls) and allows for their mental simulation. Models and simulations are guided by the logic of the figurative structures (for example, when a material is a container for cold, elements of the story must follow the logic of the CONTAINER schema).

Overall, the story suggests that there is a *force of nature* (made vivid as an agent) having properties of *quantity* (size), *intensity* (coldness, temperature), and *power*. (Originally, the idea of a simple basic gestalt of the type I call FORCE was suggested by macroscopic physical science; see Fuchs (2006, 2007, 2011) for more details.[7]) Cold can accumulate, it can flow, it can be hindered, it causes other phenomena, and it can be balanced (fought) by heat; imbalance between the intensity of cold in different places lets cold flow. Accumulation of cold makes a given material colder.

Table 1.  Metaphors for cold in a Winter Story

| Metaphors | Linguistic metaphoric expressions |
| --- | --- |
| COLD IS A THERMAL LANDSCAPE | And it got colder and colder as the winter grew stronger. The temperature *fell* and fell. When it had become terribly cold and the temperature was very, *very low* … |
| COLD IS A (FLUID) (MOVING) SUBSTANCE/OBJECT | The cold of winter *found its way into* the area and *spread out*. It could *flow* into the hollow … it could *collect* there … The cold could *sneak* in through tiny cracks between walls and windows … |
| COLD IS A POWERFUL AGENT (MOVING FORCE) | The cold of winter knew a good place where *it could do its job of making everything and everybody cold* … It went into the snow lying on the ground to *make it very cold* as well and this made the snow drier and harder to work with. The fires in the furnaces *had to work very hard to fight* the cold. |

Clearly, our story frames a scene. The question remains how properly scientific this framing can be seen to be. Sadi Carnot's verbal rendering of his idea of the role of heat in steam engines will now be used to show that the conceptual structure in the Winter Story is fundamentally the same as that found in his scientific text (Fuchs, 2010).

## Carnot's Power of Heat

Carnot's model of the power of heat serves as a prime example of the kind of metaphoric structures that are common to our perception of forces of nature. Carnot described the operation of a heat engine as follows (Carnot, 1824):

> Everyone knows that heat can produce motion. That it possesses vast motive-power no one can doubt, in these days when the steam-engine is everywhere so well known. (p. 3) The production of motive power is then due in steam-engines not to an actual consumption of caloric, but to its transportation from a warm body to a cold body [ … ] (p. 7) According to established principles at the present time, we can compare with sufficient accuracy the motive power of heat to that of a fall of water [ … ]. The motive power of a fall of water depends on its height and on the quantity of the liquid; the motive power of heat depends also on the quantity of caloric used, and on what may be termed, on what in fact we will call, the height of its fall, that is to say, the difference of temperature of the bodies between which the exchange of caloric is made. (p. 15)

This linguistically beautiful example reflects in a compact manner, the figures of mind we have seen operating in the Winter Story—heat is a force of nature in the sense described earlier. Nature or machines create a thermal tension (temperature difference) that lets quantities of heat (caloric) flow like water in a waterfall. As it turns out, the power of heat, that is, the measure of its causative force, results from the measures of tension and quantity[8] combined:

$$\text{Power of heat} = \text{Flow of caloric} \cdot \text{Thermal tension.}$$

The conceptualization of the gestalt of heat in terms of intensity (tension: understood metaphorically by the projection of the SCALE schema), quantity (a FLUID SUBSTANCE metaphor is used to conceptualize such quantities), and power (metaphoric projection of the gestalt of direct manipulation, see Lakoff and Johnson, 1980) is the starting point for the construction of modern (continuum) thermodynamics (Fuchs, 1996/2010). Naturally, we recruit additional schematic and metaphoric structures to understand the properties of a force such as heat. Aspects of fluid substance use projections of schemas such as FLOW, CREATION, and CONTAINER, whereas flow uses ENABLING or RESISTANCE. Clearly, there are a fair number of fine-grained elements of a metaphoric network to be found in our conceptualization of forces.

*Summary*

Carnot's text is a non-central member of the narrative category (even though the first paragraph quoted above hints at a possible full story): it basically recounts the bottom-up view of the force of nature called *heat*. We may think of the passage as a narrativized explanation (a text type intended as an explanation but written in a form that includes elements of narrative; see the earlier text). Still, it is an example of narrative framing of natural (and technical) scenes: it supports narrative perception (the enlisting of narrative intelligence). It conjures up images of the agent called heat. No matter how short the description, it transports us into a storyworld where the character of the force of heat is described clearly while semi-formally.

Our Winter Story, on the other hand, comes very close to what we call a central member of the category of narrative according to the four requirements listed earlier (Herman, 2009). It embeds natural agents in a story, frames a natural scene, and describes how the agents act or suffer in this world. Importantly, agents' characters are described metaphorically in terms of the same conceptual structures that are found again in corresponding scientific accounts. Heat (rather than cold) is a quantity that flows and can be stored; temperature is its potential; and the flow of heat from higher to lower potential drives other processes (remember Carnot's text and note the case of a formal model to be presented in the next section).

However, an important part of the argument that the Winter Story is a prototypical narrative rests upon our willingness to give the natural agents a character similar to sentient intentional beings. Conversely, this means that we, as interpreters, must be touched emotionally by such natural characters; in other words, it means that we must be able to get to know forces of nature emotionally (both through stories and direct natural experience) and ground our intellectual understanding in such an emotional foundation (see Endnote 6 for a brief discussion of what is involved in this issue). We will have proper stories not only due to the appearance of human (or human like) intentional agents but also when we recount the adventures of forces of nature. The argument can be summarized as follows: our encounters with nature become narratable just as our encounters with other humans do.

In short, although they have radically different origins, both examples discussed here stimulate *narrative perception* of natural scenes—they prompt the creation of natural storyworlds (remember that narrative perception is one of the senses of narrative framing). Importantly, since they have the same narrative elements (both conjure up scenes with forces of nature as agents), the storyworlds are of the *same* type (put more formally, they suggest the same scientific concepts). Conversely, this means that we can write proper stories such as the Winter Story that can be scientifically relevant (this is the sense of *narrative production* alluded to in the definition of narrative framing given in the Introduction).

## Narrative Framing in a Formal Model and Its Simulation

The notion of narrative framing is not solely dependent upon our being able to produce prototypical narratives, that is, stories for scientific purposes. The first of the senses mentioned in the description of framing presented earlier—enlisting of narrative intelligence in the perception of nature and semiotic products—will still be at work even if we have a formal text that is not at all story-like. In order to demonstrate this, I will discuss the example of a formal mathematical model of electric heating of water in a teakettle (see Figure 1) and its simulation(s). To simplify matters, it will be presented in the uniform dynamical systems version of a continuum physics model (Endnote 1). In contrast to an example presented in everyday language, we have here the opportunity to read figures of mind from the form of equations (remember that I am claiming that narrative perception can still work in the case of a mathematical text).[9]

*The Model*

Imagine some water in an electric teakettle. When the electricity is turned on, the water will get hotter over time and, because of the loss of heat through the kettle

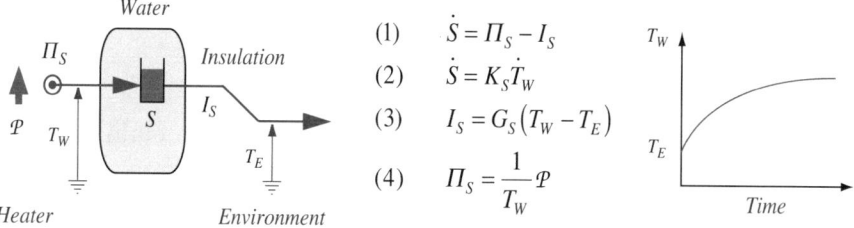

$$(1) \quad \dot{S} = \Pi_S - I_S$$

$$(2) \quad \dot{S} = K_S \dot{T}_W$$

$$(3) \quad I_S = G_S(T_W - T_E)$$

$$(4) \quad \Pi_S = \frac{1}{T_W} \mathcal{P}$$

Figure 1.   Process diagram (left), equations, and simulation of a dynamical model of the heating of water in a teakettle. The quantities shown are (electric and thermal) power ($\mathcal{P}$), entropy ($S$), flow of entropy ($I_S$), production rate of entropy ($\Pi_S$), temperature of water ($T_W$), temperature of environment ($T_E$), entropy capacitance of water ($K_S$) and conductance for flow of entropy through the wall of the kettle ($G_S$).

wall, its temperature will reach a steady state at a level that depends upon the power of heating (Figure 1, right). The mathematical model (Figure 1, center) makes use of the law of balance of entropy for heater and water (entropy is produced in the heater and communicated to the water; the water stores entropy and emits it to the environment; Figure 1, Equation (1)); the constitutive relation between entropy stored and water temperature (Equation (2)); the constitutive relation between entropy loss to the environment and temperature difference between water and environment (Equation (3)); and the relation between electric power, entropy production rate, and water temperature (Equation (4)). In addition, an initial condition for the entropy of the body of water, electric power, and constitutive quantities (capacitance and conductance) need to be specified (not shown here).

## Simulation

Stripped to its bare bones, a model such as the one presented earlier is a system of equations (expressions or statements). A simulation leads to quite a different semiotic product—it is both an activity and a result that has lately been related to storytelling and stories (see the later text).

Importantly, time only makes its proper appearance in a simulation. Even though time is written in the equations of dynamical models (see Figure 1), only a simulation involves evolution (tracing paths) through time. Only a simulation shows the full meaning and importance of the variables that are related by the equations of the model. Simulations show us the full range of possibilities of behavior (in the virtual world) inherent in the model.

Furthermore, the model does not specify every single one of the elements necessary for its simulation. In particular, initial values and parameters are not given (prescribed) by the model. Specifying them involves using the model for particular purposes, an act that cannot be defined as part of a theory to which the model belongs. To create this definition, a practitioner has to embed the model in the world. This act involves a mental attitude that goes beyond what a particular theory provides to us: it is a narrative act (see Morgan, 2001, 2012; and in the following).

## Metaphoric and Narrative Interpretation of Model and Simulation

When we read the equations we can get a feeling for the figures of mind they reflect.[10] Equation (1) is the law of balance of entropy. It suggests that entropy is imagined as a *fluid quantity* that is contained in bodies and whose amount can change ($dS/dt$) due to flow ($I_S$) and production ($\Pi_S$).[11] When the amount of entropy in the water changes, the temperature of the water ($T_W$) changes in parallel according to the properties of the container ($K_S$: the entropy capacitance of the water; see Equation (2)). Equation (3) tells us that entropy flows due to the *thermal tension* between the hotter (water) and colder (air) bodies. Temperature is a *level* (potential) whose difference is felt as a tension. The insulation of the kettle lets entropy through (or obstructs the flow; $G_S$ is the conductance for the flow of entropy from water to air).

Finally, we make use of the notion of *power* of a force of nature. This calls into existence at least one more force of nature, in this case electricity (see the left of the diagram in Figure 1). Electricity is the agent responsible for the production of entropy (the patient). Power is the measure of their interaction: the first agent (electricity) makes energy available at a certain rate; the energy is used to produce entropy in the electric heater. The measure of power of the first agent is equal to the causative power influencing the patient that is equal to the measures of the thermal tension in this process (difference of temperatures between heater and absolute zero) and quantity of entropy conjoined.

When we (mentally) simulate the model, or if we recount a simulation in natural language, we are setting up a concrete scenario in the world of the figures of mind (in the storyworld) of the model and then follow the agent(s) through a sequence of events.

*Summary*

Clearly, however they present themselves, a mathematical model and a graphical or tabular representation of a simulation are not stories, not even a more peripheral version of a narrative. The process diagram on the left in Figure 1—while representing concepts in the form of visual metaphors—is not a story either.

Nevertheless, narrative framing can (and should) still happen even though neither a story nor forces of nature nor metaphors may be directly visible in the semiotic product.[12] The argument rests upon the assumption that we, as readers and learners, can *perceive* narrative structures in the model and its use—we enlist narrative (and metaphoric) understanding in interpreting the model. Clearly, we have a very similar network of embodied conceptual relations as in the example of the Winter Story. Heat is a force of nature structured in terms of metaphors like the ones used before, and it is understood in terms of its role in simulations of the model. The model itself represents a characterization of a storyworld (including a specification of the character of the agents), and a simulation is like telling a particular story with this model.[13]

## Narrative Framing: Stories and Storyworlds, Simulations and Models

Two lines of research concerning science and narrative in the fields of computational science and economics will be seen as greatly enhancing our investigation of narrative as a central component of the production and reception of science. In the previous sections, I have referred to a relation between scientific models and storyworlds. This point will now be discussed in more detail leading us to a model of the relation between embodied conceptual structures and semiotic products and acts in the realm of science (see Figure 2 for a graphical representation of the model).[14]

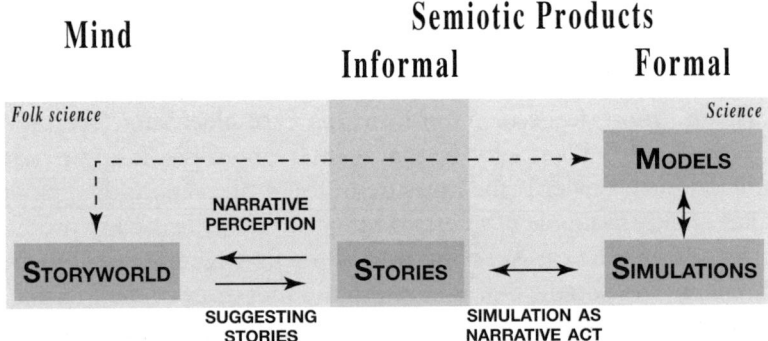

Figure 2. A model of the relation between formal models and storyworlds. The relation is mediated by simulating-as-storytelling-with-models and stories-transporting-us-into-storyworlds. Note that the relation between models and storyworlds is indirect—it emerges from the other relations.

### Growing Solutions in Computational Science as Storytelling

A particularly interesting case of *science as historicizing narrative* has been described by Wise (2011). Wise contrasts the traditional mode of explanation as deduction from differential equations with explanation through simulation. He argues that narratives that accompany simulations are historical in kind: they (the narratives) *explain* natural phenomena by growing (developing) them rather than by referring them to general laws. This is a phenomenon well known to those working in computational fields of science: explanations (of the behavior of systems) grow from many simulations; a picture emerges from a vast number of trajectories followed rather than from a single analytic solution of a set of equations.

### Narrative Embedding

The following application of narrative in social science has been described by Morgan (2001). I consider this paper on economic modeling one of the most important (not the least for the natural sciences) for showing how far the theory of narrative in science has come (for an extended and more recent discussion, see Morgan, 2012). Morgan demonstrates that *using models* and relating them to the world is a *narrative activity*. How to make use of a model or what to look for in a simulation are aspects that are not covered by the model or its underlying theory. Posing questions that lead to the definition of parameters and initial values for subsequent simulation is a narrative act.

Both Morgan's and Wise's investigations point to a strong and rather direct relation between narrative and simulation (Morgan: telling stories with models)—this will prove central to my model of the role of narrative framing presented in the later text. Note that researchers seem to draw a line when it comes to using narrative for helping frame natural and technical scenes in the sense discussed in the previous sections. It is not assumed that narratives frame scenes in a manner that would lead us to formulate concepts and models. Models are derived from theory in a classical manner and theory is given—it exists prior to modeling and the subsequent narrative act of embedding models in the world. Storytelling does not lead to theory.

*Stories Suggest Concepts and Models*

This is clearly too narrow an interpretation of the research cited (Morgan, 2001; Wise, 2011)—it would be strange if there were no feedback from the act of simulating as storytelling to the construction of models. Since 2001, Morgan has extended her work on narrative and modeling (see Morgan, 2012). She writes that:

> after my initial paper on stories (2001) [ . . . ] I tried to push further the ideas about narrative in the way that economists work with models. I came to distinguish between two things. One was the way stories form the identity of a model—the stories that can be told by working with a model (in simulations [ . . . ]) are the way that economists fully understand what kind of a model it is. Second, there are stories that are developed to map/match those models onto features of the world. (Private communication, 2014)

By using the notion of storyworld, we can explain how stories lead to the construction of scientific models and how such models can be read narratively (meaning that they lead to the construction of storyworlds; see Figure 2); note that the interaction between models and storyworlds is assumed to be indirect. I accept Morgan's (and Wise's) idea that simulating models means telling stories (with these models) as a central feature of my proposal. If we further accept that scientific stories (such as the Winter Story or Carnot's text) let us create scientific storyworlds—that is, conceptual structures (frames) that contain the seeds of scientific ideas—we can consider models the formal counterparts of storyworlds. In other words, stories of forces of nature relate to storyworlds as do simulations to models. Stories that frame natural scenes transport us into (natural) storyworlds where we create (narrative) understanding of forces of nature.

This closes the circle: models suggest stories and stories suggest models (via their storyworlds). Stories have the power to propose concepts and models and, therefore, elements of theory.[15] The stories recounting our encounters with nature contain the elements that become building blocks of theoretical knowledge. Framing of natural scenes originates in narrative perception of nature whose products find their way into our formal scientific models and theories.

## Conclusion: Narrative Framing and Science Education Research

The notion of narrative framing of natural and technical scenes suggests a number of consequences for science, science learning, and science education research. In the practice of modeling and simulation, narrative is seen as a methodological tool that helps with conceptualizing new situations. Science learning, on the other hand, can profit from an application of narrative that goes beyond creating affect and historical and social background and even beyond using narrative for explaining particular natural–historical events. In order to better understand novel uses of narrative in science learning, previous research in this field will be described briefly in the later text.

As for science education research, the present study suggests that there is more than just a superficial link between everyday and scientific forms of thought—we can expect linguistic studies to shed fresh light upon questions in conceptual change research (see Amin, 2009). Finally, if we accept that scientific thinking is deeply embedded in some forms of narratives (such as stories of forces of nature), we can extend applications of cognitive linguistics and the concept of embodied cognition to semiotic products that are larger than single utterances—we can probe students' understanding of science in situations that involve more than a single concept. It will become possible to investigate large-scale modeling activities in science learning (Fuchs, 2006; Hestenes, 2006).

The discussion in this paper has been based upon developments in modern continuum physics, investigations of narrative in computational science and economics, and modern narratology, all of which lie somewhat outside the scope of current research in science education. Therefore, it is time to embed the theme of this paper into wider concerns, particularly those of studies of narrative in science learning and of conceptual metaphor and embodied cognition. Finally, I will give a view to research into applications of narrative framing.

## Narrative in Science Education

Storytelling in science classrooms has attracted considerable interest in recent years. We are confident that stories can be used to create affective environments for social–historical contexts of science to engage and motivate learners; here, storytelling remains *extrinsic* to science. Some applications, discussed under the heading of *narrative explanation*, introduce the concept of narratives *intrinsic* to science (the distinction between extrinsic and intrinsic forms of narratives has been made by Norris et al., 2005).

How does the idea of narrative framing fit in with previous work on narrative in science and science education? Since narrative understanding is most easily associated with the non-paradigmatic (see Bruner, 1987, 1990, who seems to assume that there is a dichotomy between narrative and paradigmatic modes of understanding), it is not surprising that applications of narrative in science have been studied mostly for *extrinsic* cases. For instance, there is the grand narrative of meaning of science and of scientific knowledge (Lyotard, 1979/1984; science has a meaning for understanding human culture and the human condition). Then, there are less grand examples such as when we say that the claims of science or a part thereof (such as of thermodynamics) are a 'story.' Arnold and Millar (1996, p. 251) tell us that:

> [t]he scientific 'story' about thermal phenomena then says that, if two objects at different temperatures are placed in thermal contact, heat will spontaneously flow from the one at higher temperature to the one at lower temperature. [ . . . ] the 'story' must be accepted *as a piece*; it only makes sense as a complete 'story'.

So, there is a 'scientific story' as opposed to a possibly non-scientific one, and it is a story because it connects conceptual elements into a whole rather than leaving them as a more or less loosely packed conglomerate of statements (laws, etc.).

Stories *about* science have been created and investigated for a range of applications. An important goal of authors of such stories is to create affect (Egan, 1986); Spoel, Goforth, Cheu and Pearson (2008) discuss an example of apocalyptic narrative explanation in the field of climate change meant to engage citizens. Kubli (2001, 2005) shows how storytelling can be employed very generally for creating an environment conducive to learning about science. Stories about science and technology can create historical and social context (Klassen, 2006; Levinson, 2006; Metz, Klassen, McMillan, Clough, & Olson, 2007). Learning of science is supported by creating reading materials where expository texts are blended with narrative elements (Avraamidou and Osborn, 2009). Finally, narrative has been investigated for promoting conceptual change (Klassen, 2010), and providing background for learning about the process of science (Bruner, 1996, p. 126).

In their study of intrinsic uses of narrative, Norris et al. (2005) present a thorough discussion of the concept of *narrative explanation*. With regard to the phenomenon of explanation, the authors argue against a narrow reading of *explanation* only in terms of the *deductive* or *deductive-nomological model* (the covering law model; see, for example, Hempel and Oppenheim, 1948). Briefly stated, according to this model, an explanation of a phenomenon refers to the general laws and initial condition from which a solution can be derived that reflects the observations. (In this sense, the initial value problem formulated in Figure 1 is an explanation of the phenomenon of electric heating of water in the teakettle.) Norris et al. (2005) show how the deductive reading of explanation cannot do justice to a vast range of science. Where science treats either *singular events* (a meteorite striking the Earth 65 million years ago) or *historical events in nature* (stellar evolution, the development of a particular ecosystem), strict deduction fails. In the end, only narratives of the (special, singular, historical) events can be produced. Such stories properly and sufficiently *explain* what we want to know.

Actually, this use of stories is related to simulating models as explained earlier. Assuming that there is a model behind it, telling a story of the demise of the dinosaurs (Norris et al., 2005) is similar to how economists tell stories in their use of models (Morgan, 2001, 2012). So, there is a direct link between some of the research in narrative in science learning and the present theme. And even though extrinsic uses of narrative seem to be far from what I have discussed in this paper, note how important its concern with basic aspects of narrative—such as the elicitation of emotion and affect—is for our subject here (see earlier and Endnote 6).

*Embodied Cognition in Science Learning*

Much of the present paper has been motivated by cognitive linguistics and the concept of embodied cognition (see, for example, the characterization of forces of nature in terms of an embodied conceptual network described earlier). In science education, conceptual metaphor theory has been employed to inform us about common sense conceptualizations used by learners and professional scientists alike (Amin, 2001, 2009; Amin, Jeppsson, Haglund, & Stromdahl, 2012; Brookes and Etkina, 2007; Fuchs, 2006, 2007; Jeppsson, Haglund, Amin, & Stromdahl, 2013; Lancor, 2014a, 2014b). Linguistic investigations motivated by the concept of embodied cognition help us understand students' reasoning in science and cast light upon figurative structures of the human mind at the same time. We become more sensitive to everyday forms of reasoning that should help us become better teachers. Moreover, we understand better in what way common sense reasoning is a productive resource in learning.

Language is an agent of active learning and conceptual change, not just a tool for probing the mind. A quote from Amin (2009, p. 166) neatly summarizes the role of feedback from language and semiotic products to the mind: 'It is suggested that the appropriation of construals implicit in language and the metaphorical nature of our understanding of many concepts pervasively reflected in language, together, are likely to constitute important sources of conceptual change.' This parallels the discussion of narrative framing presented with the examples in earlier sections. I consider narrative perception of natural scenes prompted by texts (or more generally, semiotic products) as a process analogous to what Amin calls 'the appropriation of construals'; this indicates that such texts should be seen as important sources of learning, including learning how to use language for understanding.

An early investigation by Andersson (1986) shows how cognitive linguistics and the theory of conceptual metaphor (Lakoff and Johnson, 1980) can be used to unify observations of preconceptions of learners in science. Furthermore, Bliss (2008) has come up with suggestions for understanding student reasoning as based upon the role of (image) schemas formed by our perception of the natural world. Her proposal is also an attempt at unifying observations of everyday understanding of macroscopic physical phenomena. Andersson's discussion of the *experiential gestalt of causation* and Bliss' schemas are important components of my concept of forces of nature. As I have discussed earlier, the conceptual structure of forces is made of metaphoric projections of image schemas many of which bear resemblance with Bliss' schemas. Moreover, the aspect of power is structured very similar to what we learn from the gestalt of causation.

Embedding these various lines of research into investigations related to narratology widens the field of inquiry. This paper is an attempt at bringing together cognitive narratology with cognitive linguistics and its applications in science learning.

*Outlook: Researching Applications of Narrative Framing*

Narrative framing has been explored, and continues to be explored, for a number of applications. The theory of uniform dynamical thermal systems has been produced explicitly upon framing thermal phenomena in terms of forces of nature (Fuchs, 1996/2010). A course on physics as a systems science for engineering students using this approach has been taught for over 10 years (Dumont et al., 2014). A story approach to mechanics is being developed and investigated in an Industrial Educational Laboratory at Ducati in Bologna, Italy (Ascari, Corni, Corridoni, Anna, &

Savino, 2013). An art student at Zurich University of the Arts has produced an animated movie for her bachelor's thesis that transforms stories of forces of nature into an animation creating and using visual metaphors (Deichmann, 2014). Based on her approach, plays are designed for workshops on science and technology taught at VW's Autostadt campus. Most importantly, stories of forces of nature have been written by Fuchs (2011–2014, Private communication) and Fuchs (2013a, 2013b) for inclusion in a narratively shaped primary school curriculum and teacher training in Italy (Corni, 2013, 2014; Corni et al., 2013); here, the entire curriculum takes a cue from narrative science education.

Much research remains to be done on narrative framing for applications in science, not just regarding details but fundamental questions as well. Implicit in my arguments is a model of embodied conceptual structures (including narrative) created by organisms interacting with their environment that still needs to be worked out. We need to research what it means when student teachers learn science in a narrative approach and we want to understand better what happens when children are exposed to stories of forces of nature and directly to nature—what storyworlds do they construct, what are the results of direct perception of events in nature corresponding in scale to what we may call stories, and how do these results relate to storyworlds? Forces of nature and narrative framing are expected to be important in this endeavor.

## Acknowledgements

I would like to thank Federico Corni, Alessandro Ascari, Enrico Giliberti, and Elisabeth Dumont for valuable discussions of the use of stories in teacher training, in school, and in industrial settings, and for allowing me be a part of their research projects. Moreover, I greatly appreciate the eminently constructive feedback of a reviewer and the guest editors on a first version of this paper. Finally, I would like to thank my wife, Robin Fuchs, for helping me design stories of forces of nature for primary education.

## Notes

1. As far as science is concerned, the present analysis is limited to macroscopic physical science—essentially in the form of continuum physics and its simple derivative of spatially uniform dynamical systems. While the literature on continuum physics is vast, it is by nature very technical. Some of the books on continuum physics in general (Eringen, 1971–1976; Truesdell & Noll, 1965; Truesdell & Toupin, 1960) and on continuum thermodynamics in particular (Jou, Casas-Vazquez, Lebon, 1996; Müller, 1985; Truesdell, 1984) may be of use for those interested in an overview and technical aspects. The most accessible of these may be a text on a modern dynamical theory of heat (Fuchs, 2010). Here is a brief description of the basic structure of continuum physics (Fuchs, 2010, p. 9). Modern continuum physics presents us with a unified approach to macroscopic physical systems and processes taking the following form. First, we have to agree upon which physical quantities we are going to use as the fundamental or *primitive* ones. On their basis, other quantities are defined and laws expressed. Second, there are the fundamental *laws of balance* of the quantities that are exchanged or created in processes, such as momentum, charge, entropy, or amount of substance; I call these quantities *fluidlike*. Third,

there are potentials or potential differences that are visualized as driving forces for the processes undergone by the fluidlike quantities. Fourth, we need particular laws governing the behavior of, or distinguishing between, different bodies; these laws are called *constitutive relations*. Constitutive laws relate the basic fluidlike quantities to potentials or potential differences. Last but not least, we need a means of relating different types of physical phenomena to each other. The tool that permits us to do this is energy. We use the *energy principle*, that is, the law that expresses our belief that there is a conserved quantity appearing in all phenomena that has a particular relationship with each type of processes.

2. Figures of mind: objects of (in) mind created by figurative thought (e.g. a metaphor) or by perception (a perceptual gestalt or shape). Short for *figurative structures of mind*; to be distinguished from figurative structures outside of mind and structures of mind that are not figurative. (On the notion of figurative thought and related concepts, see Gibbs, 1994.) In using the term figure of mind, I take a cue from *figure of speech*. Traditionally, such as in rhetoric, examples of figures of speech are metaphor, simile, hyperbole, synechdoche, etc. Since cognitive linguistics insists that figurative language reflects a deeper (i.e. mental) phenomenon, the term *figure of mind* is introduced as the (mental, cognitive) counterpart of figure of speech. It relates not to semiotic products but to mental products, objects, or structures. Sometimes I use the term more broadly to include the (direct) products of perception, that is, *gestalts* or *shapes* (the latter term is from Arnheim, 1969). Arnheim's notion of 'visual thinking' suggests that (visual) perception is a form of thinking leading to structures of (figurative) thought. When the distinction between shapes or gestalts and the products of projection of structure from such gestalts onto target domains (as in metaphor) is important, I will use figures of mind only for the latter objects. See Section 2.1 for further details.

3. Note that I use the term *force* not in the sense of mechanics proper but in its primitive sense of phenomena that are endowed with power. Heat, wind, justice, language, pain, love, electricity, music, the market, etc. are forces or powers in this sense (music has been described as, but not named, a force by Johnson, 2007, Chap. 11). Macroscopic physical science grows from the notion of forces of nature (Fuchs, 2010).

4. Framing is used in a sense originally suggested by Fillmore's (2006) frame semantics: if we hear an utterance we construct—or, if it has been constructed before—invoke a frame, that is, a conceptual structure for understanding that utterance. The latter is said to be 'about a scene', so we can speak of framing scenes (or situations or scenarios; see also Cienki, 2007).

5. In cognitive science, it is quite common to assume that humans have a narrative mind, meaning that they understand the world narratively (see, for example, Bruner's concept of narrative construction of reality, Bruner, 1991; this is again a theme in modern narratology as a branch of cognitive science; Herman, 2013). Building on this concept, I assume that our narrative mind allows for *intelligent narrative perception*: we do not just perceive small-scale stuff as units from which larger-scale things are built but also large-scale processes and events that resemble (long) stories. This argument parallels Rudolf Arnheim's concept of the intelligence of visual perception (Arnheim, 1969) and may be recognized again in the idea of the narrativity of perception (Carr, 1991).

6. The question of emotional perception in a story of natural forces is quite central to the entire issue of using narrative in a more than peripheral (extrinsic: Norris et al., 2005) form in science and science learning. On the one hand, it has been argued quite convincingly that story and emotion go hand in hand. At the end of a good story, we should know how to feel about the events and characters (Egan, 1986). Stories give us emotional closure, not intellectual understanding (Velleman, 2003). Stories are opposed to paradigmatic thought; they are repositories of folk psychology (Bruner, 1987, 1990). On the other hand, emotion may well be the root of reason (Johnson, 2007). Therefore, accepting that a good story (a central member of the category of narrative) must make emotional perception possible and that a good scientific story must make narrative framing of forces of nature possible, we need to admit that science stories

have to open emotional access to these forces. The question of how stories of forces of nature lead to their emotional perception and, eventually, to their scientific framing, will have to be investigated in much more depth in the future. Clearly, this is just one example of the quest for understanding the relation between emotion and reason.

7.  I would like to suggest that the concept of a gestalt of force and its figurative structuring goes well beyond natural phenomena. Indeed, forces are ubiquitous creatures of the human mind; we perceive social and psychological forces in addition to forces of nature. A beautiful example of the metaphoric analysis of our understanding of music (Johnson, 2007, Chap. 11) shows not only that there are other forces, it demonstrates a particularly useful form of analysis of a phenomenon that is perceived as what I call force. Johnson shows that there are three main conceptual metaphors we use in our understanding of music: (1) music as a moving object, (2) music as a landscape in which we move, and (3) music as a moving force (the identification of these metaphors parallels the structure of substance, intensity, and power I use to conceptualize forces of nature). Another example is the perception of justice where the full conceptual structure of a force is reflected in everyday utterances (Fuchs, 2011).

8.  The power of a process is always equal to the product of a potential difference (tension) and the flow of a fluidlike quantity through this potential difference. In fluids: $P_{\text{fluid}} = \Delta p \cdot I_V$ ($p$: pressure, $I_V$: volume current); in electricity: $P_{\text{electric}} = \Delta \phi \cdot I_Q$ ($\phi$: electric potential, $I_Q$: current of electric charge); in thermodynamics: $P_{\text{thermal}} = \Delta T \cdot I_S$ ($T$: temperature, $I_S$: current of caloric (entropy)); in chemistry: $P_{\text{chemical}} = \Delta \mu \cdot I_n$ ($\mu$: chemical potential, $I_n$: current of amount of substance). All these equations can be represented in terms of visual metaphors, so-called process diagrams (Figure 1; Fuchs, 2010, Chap. 2; see also Falk, Herrmann, & Schmid, 1983).

9.  It is certainly possible to argue, as many scientists would, that equations do not invoke images. On the other hand, if we assume that our concepts are grounded in embodied cognition, we cannot escape the conclusion that we *must* find figures of mind in the equations of a model. For an important argument that equations are more than purely formal representations, see Sherin (2001). Extending the depth and details of Sherin's work to the present case is beyond the scope of this paper. For a more detailed description of the figurative structure of the partial differential equations of continuum thermodynamics, see Fuchs (2013c). See also Endnote 11.

10. For a more sophisticated example of continuum thermodynamics (thermoelectricity), see Fuchs (2014). We find the same figurative conceptual structures there as in our simpler example. The ease with which a supposedly complicated case is modeled is a witness to the power of the storyworld that is the result of a narrative approach to thermodynamics.

11. In modern (non-equilibrium and continuum) thermodynamics, the law of balance of entropy is conceptualized as the formal equivalent of the embodied notion of the balance of a fluidlike quantity analogous to charge, momentum, or amount of substance (Fuchs, 1996/2010: Introduction). Unlike charge or momentum, entropy satisfies only half a conservation law, and amount of substance none at all. Entropy satisfies the embodied notion of caloric suitably extended by the requirement that caloric is generated in irreversible processes. On the concept of caloric in the modern theories of thermodynamics, see Callendar (1911), Job (1972), Falk (1985), Fuchs (1986, 1987a, 1987b, 1996/2010) and Mareš et al. (2008). For a contribution to the debate of historical issues, see Kuhn (1955). Let me stress here that I personally believe that changing the use of terminology from entropy to caloric would be essential on two important grounds. First, psychologically speaking, the word entropy does not convey any useful embodied image, certainly not for macroscopic models of thermal processes. This word will serve no other purpose than to confuse a child and send an adult layperson onto a search into esoteric land or for microscopic disorder. Second, for scientific reasons, it is paramount that we understand the difference between macroscopic and microscopic models (and accept that microscopic models do not ground macroscopic ones). Using two terms, caloric

and (logarithm of) number of possible configurations or states, for macroscopic and microscopic models, respectively, can make this distinction plain. I have refrained in this section from using the term caloric in order to conform to the tradition.

12. Note that we could easily construct a word model for the situation analyzed here. Its form will depend upon the audience this is intended for (engineering students or young children, for instance). Depending upon the circumstances, the semiotic product could be a peripheral member of the category of narrative (a narrativized description or an explanatory narrative) or a central member, that is, a proper story (see Section 'Theory of narrative').

13. Few practitioners and teachers of science will spontaneously interpret the mathematical model and its simulation in terms of narrative and story (in economics and in computational science, the practice seems to be different, though; see Section 'Narrative Framing'). However, this does not change the validity of what has been said. Rather, it compels us to rethink education in science. If we are not taught so, we will not develop a 'narrative eye' for what we see and do. We will simply follow the tradition and accept a system of equations as the only true and objective but otherwise meaningless collection of signs reflecting nature directly 'as it is'.

14. In this model, the relation between mind and (natural) world is missing. In this paper, I have made use of an assumption regarding this relation in the form of the claim that enlisting narrative intelligence (or the act of narrative perception) also refers to our perception of nature (see Section Introduction and Endnote 5).

15. Here is an example for how narrative thinking influences (the production of) science. Note how important figures of mind are for the construction of a theory. In continuum physics, the basic structure of the gestalt of forces guides the choice of *primitive quantities* (Endnote 1). In all fields (fluids, electricity and magnetism, heat, chemical substances, translational and rotational motion), this choice takes the same form: primitive quantities for a theory are (1) the potential and (2) a fluidlike quantity, and directly related quantities such as stored amount, current, production rate, and potential difference. The primitives of modern thermodynamics are temperature and temperature difference, entropy (caloric), current and production rate of entropy. This choice is fundamentally important (Fuchs, 1986, 1987a, 1987b, 2010, pp. 1–13).

## References

Amin, T. G. (2001). A cognitive linguistics approach to the layperson's understanding of thermal phenomena. In A. Cienki, B. J. Luka, & M. B. Smith (Eds.), *Conceptual and discourse factors in linguistic structure* (pp. 27–44). Stanford, CA: CSLI Publications.

Amin, T. G. (2009). Conceptual metaphor meets conceptual change. *Human Development, 52*(3), 165–197.

Amin, T. G., Jeppsson, F., Haglund, J., & Strömdahl, H. (2012). Arrow of time: Metaphorical construals of entropy and the second law of thermodynamics. *Science Education, 95*(5), 818–848.

Andersson, B. (1986). The experiential gestalt of causation: A common core to pupils' preconceptions in science. *European Journal of Science Education, 8*(2), 155–171.

Arnheim, R. (1969). *Visual thinking*. Berkeley, CA: University of California Press.

Arnold, M., & Millar, R. (1996). Learning the scientific 'story': A case study in the teaching and learning of elementary thermodynamics. *Science Education, 80*(3), 249–281.

Ascari, A., Corni, F., Corridoni, T., D'Anna, M., & Savino, G. (2013, September). New perspective on rotation: experiments, modelling and analogies based on angular momentum as an extensive quantity. *Proceedings of the ESERA conference*, Nicosia, Cyprus.

Avraamidou, L., & Osborn, J. (2009). The role of narrative in communicating science. *International Journal of Science Education, 31*(12), 1683–1707.

Bliss, J. (2008). Commonsense reasoning about the physical world. *Studies in Science Education, 44*(2), 123–155.

Brookes, D. T., & Etkina, E. (2007). Using conceptual metaphor and functional grammar to explore how language used in physics affects student learning. *Physical Review Special Topics—Physics Education Research, 3*(010109), 1–16.

Bruner, J. (1987). *Actual minds, possible worlds.* Cambridge, MA: Harvard University Press.

Bruner, J. (1990). *Acts of meaning.* Cambridge, MA: Harvard University Press.

Bruner, J. (1991). The narrative construction of reality. *Critical Inquiry, 18*(1), 1–21.

Bruner, J. (1996). *The culture of education.* Cambridge, MA: Harvard University Press.

Callendar, H. L. (1911). The caloric theory of heat and Carnot's principle. *Proceedings of the Physical Society of London, 23*(1), 153–189.

Carnot, S. (1824). Réflexions sur la puissance motrice du feu et sur les machines propres à développer cette puissance (R. H. Thurston, Trans. and edited. Peter Smith, Gloucester, MA, 1977). Paris: Bachelier.

Carr, D. (1991). *Time, narrative, and history.* Bloomington, IN: Indiana University Press.

Chemero, A. (2009). *Radical embodied cognitive science.* Cambridge, MA: The MIT Press.

Cienki, A. (2007). Frames, idealized cognitive models, and domains. In D. Geeraerts, & H. Cuyckens (Eds.), *The oxford handbook of cognitive linguistics* (pp. 170–187). Oxford, UK: Oxford University Press.

Corni, F. (Ed.) (2013). *Le scienze nella prima educazione. Un approccio narrativo a un curricolo interdisciplinare.* Trento: Erickson.

Corni, F. (2014). Stories in physics education. In S. Burra, M. Michelini, & L. Santi (Eds.), *Frontiers of fundamental physics and physics education research* (pp. 385–396). Milano: Springer.

Corni, F., Giliberti, E., & Fuchs, H. U. (2013). Student teachers writing science stories: A case study. *Proceedings of the ESERA conference*, Cyprus.

Deichmann, M. (2014). *Im übertragenen Sinne. Metaphern und Bildvergleiche in der Wissernschaft* (Bachelor thesis). Zürcher Hochschule der Künste, Zürich. Movie to be found at http://vimeo.com/98311515

Dewey, J. (1925). Experience and nature. In J. A. Boydston (Ed.), *The later works* (pp. 1925–1953, *Vol. 1*). Carbondale, IL: Southern Illinois University Press.

Dumont, E., Fuchs, H. U., Maurer, W., & Venturini, F. (2014, August). From forces of nature to the physics of *dynamical systems. The 9th international conference on conceptual change*, Bologna, Italy.

Egan, K. (1986). *Teaching as story telling.* Chicago, IL: University of Chicago Press.

Eringen, A. C. (1971–1976). *Continuum physics (Vols. I–IV).* New York: Academic Press.

Falk, G. (1985). Entropy, a resurrection of caloric—A look at the history of thermodynamics. *European Journal of Physics, 6*(2), 108–115.

Falk, G., Herrmann, F., & Schmid, G. B. (1983). Energy forms or energy carriers? *American Journal of Physics, 51*(12), 1074–1077.

Fillmore, C. (2006). Frame semantics. In D. Geeraerts (Ed.), *Cognitive linguistics: Basic readings* (pp. 373–400). Berlin: Mouton de Gruyter.

Fuchs, H. U. (1986). A surrealistic tale of electricity. *American Journal of Physics, 54*(10), 907–909.

Fuchs, H. U. (1987a). Thermodynamics—A 'misconceived' theory. In J. D. Novak (Ed.), *Proceedings of the second international seminar on misconceptions in science and mathematics* (Vols. I–III, pp. 160–167). Ithaca, New York: Cornell University.

Fuchs, H. U. (1987b). Entropy in the teaching of introductory thermodynamics. *American Journal of Physics, 55*(3), 215–219.

Fuchs, H. U. (2006). From image schemas to dynamical models in fluids, electricity, heat, and motion. Examples, practical experience, and philosophy. *Proceedings of the 2006 GIREP conference.* University of Amsterdam.

Fuchs, H. U. (2007). *From image schemas to models in fluids, electricity, heat, and motion. An essay on physics education research.* ZHAW, Institute of Applied Mathematics and Physics www.zhaw.ch/~fusa/LITERATURE/Literature.html

Fuchs, H. U. (2010). *The dynamics of heat (2nd ed.). Graduate texts in physics*. New York: Springer (first edition: Springer, New York, 1996).

Fuchs, H. U. (2011). Force Dynamic Gestalt, metafora e pensiero scientifico. Atti del Convegno '*Innovazione nella didattica delle scienze nella scuola primaria: al crocevia fra discipline scientifiche e umanistiche*', Modena, Italy: Artestampa. English version: Force Dynamic Gestalt, Metaphor, and Scientific Thought. ZHAW, Institute of Applied Mathematics and Physics www.zhaw.ch/~fusa/LITERATURE/Literature.html

Fuchs, H. U. (2013a). Il significato in natura. In F. Corni (Ed.), *Le scienze nella prima educazione. Un approccio narrativo a un curricolo interdisciplinare* [Meaning in nature—From schematic to narrative structures of science], Erickson, Trento, Italy. ZHAW, Institute of Applied Mathematics and Physics. www.zhaw.ch/~fusa/LITERATURE/Literature.html

Fuchs, H. U. (2013b). Costruire e utilizzare storie sulle forze della natura per la comprensione primaria della scienza. In F. Corni (Ed.), *Le scienze nella prima educazione. Un approccio narrativo a un curricolo interdisciplinare* [Designing and using stories of forces of nature for primary understanding in science] Trento, Italy: Erickson. ZHAW, Institute of Applied Mathematics and Physics. www.zhaw.ch/~fusa/LITERATURE/Literature.html

Fuchs, H. U. (2013c). The narrative structure of continuum thermodynamics. *Proceedings of the ESERA conference 2013*, Cyprus.

Fuchs, H. U. (2014). A direct entropic approach to uniform and spatially continuous dynamical models of thermoelectric devices. *Energy Harvesting Systems, 1*(3–4), 253–265.

Gibbs, R. W. (1994). *The poetics of mind. Figurative thought, language, and understanding*. Cambridge: Cambridge University Press.

Gibbs, R. W. (2006). *Embodiment and cognitive science*. Cambridge: Cambridge University Press.

Gibson, J. J. (1966). *The senses considered as perceptual systems*. Boston, MA: Houghton-Mifflin.

Gibson, J. J. (1979). *The ecological approach to visual perception*. Boston, MA: Houghton-Mifflin.

Hampe, B. (2005). *From perception to meaning. Image schemas in cognitive linguistics*. Berlin: Mouton de Gruyter.

Hempel, C. G., & Oppenheim, P. (1948). Studies in the logic of explanation. *Philosophy of Science, 15*(2), 135–175.

Herman, D. (2002). *Story logic*. Lincoln, NE: University of Nebraska Press.

Herman, D. (2009). *Basic elements of narrative*. Chichester: Wiley-Blackwell.

Herman, D. (2013). *Storytelling and the sciences of mind*. Cambridge, MA: The MIT Press.

Hestenes, D. (2006). Notes for a modeling theory of science, cognition and instruction. *Proceedings of the 2006 GIREP conference*, University of Amsterdam.

James, W. (1890/1983). *The principles of psychology*. Cambridge, MA: Harvard University Press.

Jeppsson, F., Haglund, J., Amin, T. G., & Strömdahl, H. (2013). Exploring the use of conceptual metaphor in solving problems on entropy. *Journal of the Learning Sciences, 22*(1), 70–120.

Job, G. (1972). *Neudarstellung der Wärmelehre: Die Entropie als Wärme*. Frankfurt: Akademische Verlagsgesellschaft.

Johnson, M. (1987). *The body in the mind*. Chicago, IL: University of Chicago Press.

Johnson, M. (2007). *The meaning of the body*. Chicago, IL: University of Chicago Press.

Jou, D., Casas-Vazquez, J., & Lebon, G. (1996). *Extended irreversible thermodynamics* (2nd ed.). Berlin, Germany: Springer.

Klassen, S. (2006). A theoretical framework for contextual science teaching. *Interchange, 37*(1–2), 31–62.

Klassen, S. (2010). The relation of story structure to a model of conceptual change in science learning. *Science & Education, 19*(3), 305–317.

Kubli, F. (2001). Can the theory of narratives help science teachers be better storytellers? *Science & Education, 10*(6), 595–599.

Kubli, F. (2005). Science teaching as a dialogue—Bakhtin, Vygotsky and some applications in the classroom. *Science & Education, 14*(6), 501–534.

Kuhn, T. S. (1955). Carnot's version of Carnot's cycle. *American Journal of Physics*, 23(2), 91–95.

Lakoff, G. (1987). *Women, fire, and dangerous things*. Chicago, IL: University of Chicago Press.

Lakoff, G., & Johnson, M. (1980). *Metaphors we live by*. Chicago, IL: University of Chicago Press.

Lakoff, G., & Johnson, M. (1999). *Philosophy in the flesh*. New York, NY: Basic Books.

Lakoff, G., & Núñez, R. E. (2000). *Where mathematics comes from*. New York, NY: Basic Books.

Lancor, R. A. (2014a). Using metaphor theory to examine conceptions of energy in biology, chemistry, and physics. *Science & Education*, 23(6), 1245–1267.

Lancor, R. A. (2014b). Using student-generated analogies to investigate conceptions of energy: A multidisciplinary study. *International Journal of Science Education*, 36(1), 1–23.

Langacker, R. W. (1987). *Foundations of cognitive grammar, Volume I, Theoretical prerequisites*. Stanford, CA: Stanford University Press.

Langacker, R. W. (1991). *Foundations of cognitive grammar, Volume II, Descriptive application*. Stanford, CA: Stanford University Press.

Levinson, R. (2006). The use of narrative in supporting the teaching of socio-scientific issues: A study of teachers' reflections. *Interacções*, 2(4), 24–41.

Lyotard, J.-F. (1984). *The postmodern condition: A report on knowledge*. Manchester: Manchester University Press (original published in 1979).

Mareš, J. J., Hubík, P., Šesták, J., Špička, V., Krištofik, J., & Stávek, J. (2008). Phenomenological approach to the caloric theory of heat. *Thermochimica Acta*, 474(1–2), 16–24.

Metz, D., Klassen, S. McMillan, B., Clough, M., & Olson, J. (2007). Building a foundation for the use of historical narratives. *Science and Education*, 16(3–5), 313–334.

Morgan, M. S. (2001). Models, stories, and the economic world. *Journal of Economic Methodology*, 8(3), 361–384.

Morgan, M. S. (2012). *The world in the model. How economists work and think*. Cambridge: Cambridge University Press.

Müller, I. (1985). *Thermodynamics*. Boston, MA: Pitman.

Noë, A. (2004). *Action in perception*. Cambridge, MA: MIT Press.

Norris, S. P., Guilbert, S. M., Smith, M. L., Hakimelahi, S., & Phillips, L. M. (2005). A theoretical framework for narrative explanation in science. *Science Education*, 89(4), 535–563.

Rosch, E. (1973). Natural categories. *Cognitive Psychology*, 4(3), 328–50.

Sherin, B. L. (2001). How students understand physics equations. *Cognition and Instruction*, 19(4), 479–541.

Spoel, P., Goforth, D., Cheu, H., & Pearson, D. (2008). Public communication of climate change science: Engaging citizens through apocalyptic narrative explanation. *Technical Communication Quarterly*, 18(1), 49–81.

Talmy, L. (2000a). *Toward a cognitive semantics (Vol. I)*. Cambridge, MA: MIT Press.

Talmy, L. (2000b). *Toward a cognitive semantics (Vol. II)*. Cambridge, MA: MIT Press.

Thelen, E., & Smith, L. (1994). *A dynamic system approach to the development of cognition and action*. Cambridge, MA: MIT Press.

Truesdell, C. A. (1984). *Rational thermodynamics*. New York: Springer.

Truesdell, C. A., & Noll, W. (1965). The non-linear field theories of mechanics. In S. Flügge (Ed.), *Encyclopedia of physics* (v. III/3). Berlin: Springer.

Truesdell, C. A., & Toupin, R. A. (1960). The classical field theories. In S. Flügge (Ed.), *Encyclopedia of physics* (v. III/1). Berlin: Springer.

Turner, M. (1996). *The literary mind. The origins of thought and language*. New York: Oxford University Press.

Velleman, J. D. (2003). Narrative explanation. *Philosophical Review*, 112(1), 1–25.

Wise, M. N. (2011). Science as (historical) narrative. *Erkenntnis*, 75(3), 349–376.

# On the Significance of Conceptual Metaphors in Teaching and Learning Science: Commentary on Lancor; Niebert and Gropengiesser; and Fuchs

David F. Treagust[a] and Reinders Duit[b]

[a]*Science & Mathematics Education Centre, School of Science, Faculty of Science and Engineering, Curtin University, Perth, Western Australia;* [b]*IPN—Leibniz Institute for Science Education, University of Kiel, Kiel, Germany*

We are delighted to be invited for this task as our own work on conceptual change (Duit & Treagust, 2003) so far has not explicitly included ideas from the conceptual metaphor field. In other words, we see this task as a chance to revise and further develop our more 'classical' conceptual change views.

## Comments on Our Own Research Work on the Role of Analogies and Metaphors

The role of analogies and metaphors has played a significant part in our work on teaching and learning science. Reinders Duit's academic study leave at the Science and Mathematics Education Centre (SMEC) of Curtin University in 1988 was the start of that work. Duit (1991) summarised the state of literature on the role of analogies and metaphors in science teaching and learning in a review paper. Analogies were seen as comparisons of structures between two domains based on structural similarities between these domains used to initiate understanding of the key features of a concept to be learned. With regard to metaphors, we conceptualised them as analogies with particular emphasis:

> *taken as literal,* a metaphorical statement appears to be perversely asserting something to be what it is plainly known not to be. … But such 'absurdity' and 'falsity' are of the essence: in their absence, we would not have a metaphor but merely a literal utterance. (Black, 1993, p. 21)

Briefly put, a metaphor in this sense compares entities without doing so explicitly. Metaphors are comparisons where the basis of comparison must be revealed or even created by the addressee of the metaphor. A 'good' metaphor always includes some dose of surprise. Clearly, this is a different meaning compared to the conceptual metaphors discussed in the various papers of the present issue. As used in these studies, conceptual metaphors denote linguistic figures of speech based on certain embodied experiences such as the substance metaphor for the energy concept in physics.

We carried out research on the role of analogies and metaphors (in the above 'classical' sense) in various studies. We investigated, for instance, the use of analogies by teachers clarifying aspects of science content in ordinary classroom practice (Treagust, Duit, Joslin, & Lindauer, 1992), and the power of using analogies as instructional tools in science classrooms (Treagust, Venville, Harrison, & Dagher, 1996). Research on the role of analogies in guiding students to basic ideas of non-linear systems (Wilbers & Duit, 2006) questioned the classical approach as represented by Gentner (1983), especially her structure mapping idea. Namely, it turned out that it was not the similarities of certain structural features of source and target that initiated students' processes of using the analogy provided:

> Students interpret base and target domains in fundamentally different ways. Learning by analogy rests on visual perception. It traces a line of concrete visualization and abstraction by transcending the concrete in a second step. To put it into a nutshell: a student's heuristic analogy is built on mental images rather than propositionally based knowledge. (Wilbers & Duit, 2006, p. 37)

This seems to be a figure of thought whose significance is also highlighted by the conceptual metaphor approach discussed in the present volume.

Metaphors have been used as powerful tools in teacher education: Tobin (1990) and Aubusson and Webb (1992) deliberately used metaphors to encourage science teachers to revise their teaching roles. In Tobin's research, initially most teachers provided metaphors that emphasised organisation and control but later changed to metaphors such as the teacher-as-musician. Developing metaphors of teaching and learning still is an essential feature of teacher development.

In closing this section of our own background regarding the role of metaphors (in the sense of conceptual metaphors), we would like to explain how our 'classical' ideas were extended. In 2011, when Kai Niebert visited the SMEC at Curtin as a postdoctoral scholar, he introduced Treagust to the term 'conceptual metaphor' and reintroduced the work of Lakoff and Johnson and the notion of embodiment in science through metaphors. Subsequent research at Curtin showed that good instructional metaphors and analogies need embodied sources (Niebert, Marsch, & Treagust, 2012). According to philosophers, metaphors and analogies permeate all discourses, are fundamental to human thought and are not simply teaching tools. In other words, the students need to have previous experience with the metaphor or analogy—this goes beyond understanding the analog to further understand the target concept. 'One of the most important revelations is that metaphor is not merely a linguistic phenomenon but also a fundamental principle of thought and action' (Niebert et al., 2012, p. 850).

## Comments on the Three Papers

In our commentary, we start with the study that is nearest to 'classical' conceptual change studies on student conceptions research, the empirical investigation by Rachael Lancor on tertiary-level students' conceptual metaphors of energy. We continue with the work by Kai Niebert and Harald Gropengießer who illustrate how the conceptual metaphor perspective may inform planning, carrying out and evaluating instructional sequences in science and even predicting students' difficulties of understanding certain science concepts. Finally, we discuss Hans Fuchs' development of a theoretical perspective of narrative framing on the grounds of conceptual metaphor perspectives.

Briefly summarised, our analyses draw on the key features in the referring literature:

(a) Conceptual metaphors are patterns of mappings between abstract concepts and bodily based image schemas reflected in everyday language.
(b) Conceptual metaphors categorise the metaphors and analogies employed in understanding a certain topic at the level of the used source and target domain.
(c) For each of the studies presented, the underlying framework is that the image schemas (the gestalts) are structures of the embodied mind.

Paper 1: 'An Analysis of Metaphors Used by Students to Describe Energy in an Interdisciplinary General Science Course'. (Rachael Lancor)

In the abstract, Rachael Lancor summarises the major features of her paper in that when students explain "the role of energy in five contexts that frequently appear in the media; radiation, transportation, generating electricity, earthquakes, and the big bang theory, . . . they spontaneously use metaphorical language" . . . in multiple coherent ways.

Lancor (2014a, 2014b) has carried out a number of additional studies on understanding energy, drawing on the same theoretical framework (conceptual metaphor based) and the same qualitative research methods focussing on different groups of students. The findings of the present study fit well with the results of other studies in the field of teaching and learning energy (Chen et al., 2014). A particular strength is that the investigation of the use of energy in physics, chemistry and biology contexts identifies differences of students' views on energy within different disciplinary contexts. So far, most empirical research on teaching and learning energy has been carried out in the domain of physics (Duit, 2014).

Concerning the key method used to investigate understanding of energy, a conceptual metaphor perspective is explicitly adopted. However, it seems that a key feature of 'conceptual metaphors' drawing on the conceptual metaphor school, namely that cognition is deeply embedded within bodily experiences is virtually absent (or implicit) throughout the paper. This seems to be the case for the other papers by Rachael Lancor as well. If we would replace the term 'conceptual metaphor' when reading the papers by 'students' conceptions' as used in the students' (alternative) conceptions field (Duit & Treagust, 2003), this would work well. In other words, it seems that key ideas of the conceptual metaphor field are not in the foreground. The added value of the conceptual metaphor perspective as compared to existing conceptual change ideas is not explicitly elaborated. As energy has proven to be a difficult concept to be learned (Duit, 2014), the role of conceptual metaphors to guide students towards understanding of key ideas of energy such as transfer, transformation, conservation and degradation is welcomed. In brief, this is an interesting study that reveals energy ideas of students who study a little science and may represent views of educated lay people.

Paper 2: 'Understanding Starts at the Mesocosm: Conceptual metaphor as a Framework to Develop External Representations for Science Teaching'. (Kai Niebert and Harald Gropengießer)

This is a theoretically rich paper embedding key ideas of mainstream embodied cognition within theories on the role of multiple external representations in teaching and learning (Ainsworth, 2006; Treagust & Tsui, 2013), on the one hand, and evolutionary epistemology (Vollmer, 1984), on the other hand. Consistent with research in the field, the multiple external representation perspective highlights that teaching and learning may become more effective if the same topic is modelled from various perspectives. The evolutionary epistemology provides a framework to better understand the particular role of embodied cognition. Vollmer (1984) distinguishes not only macroscopic and microscopic structures but also structures between the two which he refers to as 'mesocosmic'. This is the domain of everyday life concerns and, as argued by the authors, is where understanding starts, being 'that section of the real world we cope with perceiving and acting, sensually and motorically ...' (Vollmer, 1984, p. 89). Niebert and Gropengießer argue that the framework of conceptual metaphors and Vollmer's epistemological distinctions of micro-, meso- and macrocosm can be used as diagnostic tools to predict student understanding.

Based on the outcomes from their studies, Niebert and Gropengießer put forth the argument that external representations, particularly those that involve action, visualisation or reflection of past experiences, enable a deeper understanding of the concepts investigated. This is the notion of embodied cognition: representations that involve action, visualisation or reflection of past experiences 'shed light on the embodied conceptions that shape students' conceptual understanding' (p. 34). Similarly, it is argued that 'representations that visualise an image schema and its mapping on a scientific concept ... do not provide a new experience but induce an instance of *relived* experiences' (p. 34). As useful as this notion of lived experience may be, it is not necessarily the case that every individual is able to draw on past experiences when learning from external representations, like models, and this may be a limitation of learning by embodied cognition. The paper is rich when discussing instructional sequences based on the conceptual metaphor framework in significantly challenging domains of science in school, namely cell biology, neurobiology, the greenhouse effect and the carbon cycle particularly with the container schema and the balance schema. A particularly interesting and significant facet is that according to evolutionary epistemology, the conceptual metaphors discussed developed during evolution of the human mind and hence they are particularly deeply rooted.

Paper 3: 'From Stories to Scientific Models and Back: Narrative Framing in Modern Macroscopic Physics'. (Hans Fuchs)

Narrative issues have played a significant role in science education research for a long time (Martin & Brower, 1991). The best-known examples of such an orientation are *story telling* in science classes supporting students in understanding science content and issues of the nature of science (Kubli, 2005) and *science writing* (Yore & Treagust, 2006) supporting understanding of these science issues by writing (brief) essays about science topics. Both issues are also mentioned by Hans Fuchs in the present paper. However, this paper goes far beyond these latter ideas. Actually, a rich theoretical framework of what Fuchs calls *narrative framing* is developed on the grounds of conceptual metaphor ideas that enable both the framing of modern macroscopic physics as well as teaching and learning physics. The epistemological position Fuchs holds may be illustrated by the quote: 'Physical science is figurative (it is a representation of our imagination, not the outside world)' (Fuchs, 2013).

It is interesting to learn that the idea of narrative framing leads to particular insights into the roles of models in science and science teaching and learning. Fuchs argues, for instance, that narratives and simulations allow telling stories with models. However, storytelling does not *lead* to theory. Stories may suggest conceptions and models. Hence, modelling may be seen in terms of storytelling. In other words, scientific structures have a narrative character, so learning about stories is good preparation for formal science, as is illustrated by Fuchs in the development and implementation of a primary school science curriculum in Italy. Briefly summarised, the perspective of narrative framing developed in the paper provides insights that may contribute to the further development of science education.

A minor remark concerns the use of the quantity of *heat* by Fuchs in his paper. Fuchs views the everyday use of the term *heat* as being associated with either the gestalt of the 'Force of Nature' as a whole or specifically the material-like fluid/caloric that is a component of it, as is illustrated in the Winter Story. In this argument, Fuchs does not adopt the conventional conception of thermodynamics and so he does not mention that *heat* in everyday discourse has various meanings, including the basic ideas of the physics concepts of heat, entropy and temperature (Kesidou, Duit, & Glynn, 1995).

In addition, we comment on an idea briefly discussed by Fuchs that presenting different domains of physics (such as heat and electricity) in *analogous forms* will ease understanding; the Karlsruhe Physics Course (Falk, Herrmann, & Schmid, 1983) is an example of this approach. However, only a few studies on teaching and learning the various topics of this course are available (Kesidou & Duit, 1991; Starauschek, 2001). Support for this assumption is not convincing. Research on the role of analogies and metaphors in science teaching in general has shown that these teaching and learning aids are double-edged swords (Glynn, 1991).

Finally, we comment on the complexity of the argumentation that Hans Fuchs presents in weaving together manifold theoretical perspectives. Clearly, this paper may provide a fresh view and, hence, understanding for thinking about science instruction as well as science education research. However, without reading more of his research work, the complexity of all the major points presented is difficult to fully understand. Consequently, we believe that these rich and complex ideas need further simplification to inform science teachers how they might change their ways of thinking about science teaching and their teaching behaviours.

## Concluding Remarks

The three papers discussed above cover a wide range of studies in the spirit of conceptual metaphors—ranging from a study somewhat similar to 'classical' conceptual change, to a teacher professional development approach informed by conceptual metaphor ideas and finally to an elaborate theoretical perspective of 'narrative framing'. In other words, there is much information on 'conceptual metaphors in action' in science teaching and learning in these three papers and in the literature on conceptual metaphor in general. Each of the papers' authors use the term *conceptual metaphor* in a different way, with each supported by a different theoretical perspective: for Lancor, the framework of metaphor theory; for Fuchs, the theory of narrative framing and for Niebert and Gropengießer, experientialism as the theory of embodied cognition, with an emphasis on the mesocosm, the realm of lived experience.

We argue, based on our review of these three papers, that there is a need for more precise and operationally defined use of the term conceptual metaphor or at least for the authors to identify which kind of conceptual metaphor is being used in their studies. Fuchs (2013) has clarified three different uses of the word metaphor ('at least in the metaphor theory of cognitive linguistics', p. 5).

- Special linguistics metaphoric expressions—in narratives/stories.
- Metaphors proper—that have a conceptual structure.
- Metaphor as a process of metaphoric projection—from a source to a target domain.

Consequently, we believe that in the field of research dealing with conceptual metaphors there needs to be a clear definition (or explication) of what this term denotes when used in a particular study. Surely, cognition has to do with certain features of the human body that allow certain experiences, but which features are in the foreground? Which senses are involved? Are there psycho-motor actions of various kinds?

In addition, it seems to us that research on how the conceptual metaphor approach may become part of normal instructional practice is needed. In particular, it is necessary to theoretically elaborate the added value of the conceptual metaphor approach as compared to the so far dominating conceptual change view of teacher professional development approaches. The papers argue that the connection between everyday thinking and scientific thinking is not superficial but is deeply embedded in action, visualisation or reflection of past experiences or narratives. So, how can these ideas of metaphorical concepts be brought to the science classroom so that teachers can use the ideas effectively and efficiently? The three papers do provide examples, so it is evident that these researchers can use conceptual metaphors in their teaching. Lancor uses metaphors in teaching energy to students in university general science courses; Fuchs has designed and discusses the primary school curriculum; and Niebert and Gropengießer illustrate many lessons in various domains of secondary school science.

Finally, we would like to add a thought on how the Model of Educational Reconstruction (Duit, Gropengießer, Kattmann, Komorek, & Parchmann, 2012), which is theoretically based on classical conceptual change views, may be further developed. The process of educational reconstruction of science content for teaching and learning this content to certain groups of students may certainly gain significance if 'metaphorical' issues as those provided by Fuchs (2013) are taken into account. The basically metaphorical nature of science concepts and principles may lead to a 'content structure for instruction' that more adequately addresses views of the nature of science developed from the perspectives outlined in the present volume.

## Disclosure Statement

No potential conflict of interest was reported by the authors.

## References

Ainsworth, S. (2006). DeFT: A conceptual framework for considering learning within multiple representations. *Learning & Instruction, 16*(3), 183–198.

Aubusson, P., & Webb, C. (1992). Teacher beliefs about learning and teaching in primary science and technology. *Research in Science Education, 22*, 20–29.

Black, M. (1993). More about metaphor. In A. Ortony (Ed.), *Metaphor and thought* (pp. 19–43). Cambridge: Cambridge University Press.

Chen, R. F., Eisenkraft, A., Fortus, D., Krajcik, J., Neumann, K., Nordine, J., & Scheff, A. (2014). *Teaching and learning of energy in K-12 education.* Dordrecht: Springer.

Duit, R. (1991). On the role of analogies and metaphors in learning science. *Science Education, 75*(6), 649–672.

Duit, R. (2014). Teaching and learning the physics energy concept. In R. F. Chen, A. Eisenkraft, D. Fortus, J. Krajcik, K. Neumann, J. C. Nordine, & A. Scheff (Eds.), *Teaching and learning of energy in K-12 education* (pp. 67–85). Dordrecht: Springer.

Duit, R., Gropengießer, H., Kattmann, U., Komorek, M., & Parchmann, I. (2012). The model of educational reconstruction: A framework for improving teaching and learning science. In D.

Jorde & J. Dillon (Eds.), *The world of science education: Science education research and practice in Europe* (pp. 13–37). Rotterdam: Sense Publishers.

Duit, R., & Treagust, D. F. (2003). Conceptual change: A powerful framework for improving science teaching and learning. *International Journal of Science Education, 25*, 671–688.

Falk, G., Herrmann, F., & Schmid, G. B. (1983). Energy forms or energy carriers? *American Journal of Physics, 51*(12), 1074–1076.

Fuchs, H. U. (2013, September). From image schemas to narrative structures in science. A contribution to the symposium on conceptual metaphors and embodied cognition in science learning. Paper presented at the Biannual Meeting of the European Science Education Research Association (ESERA), Nicosia, Cyprus.

Gentner, D. (1983). Structure-mapping: A theoretical framework for analogy. *Cognitive Science, 7*, 155–170.

Glynn, S. M. (1991). Explaining science concepts: A teaching-with-analogies model. In S. M. Glynn, R. H. Yeany, & B. K. Britton (Eds.), *The psychology of learning science* (pp. 219–240). Hillsdale, NJ: Erlbaum.

Kesidou, S., & Duit, R. (1991). Wärme, Energie, Irreversibilität – Schülervorstellungen im herkömmlichen Unterricht und im Karlsruher Ansatz [Heat, energy, irreversibility – Student conceptions in traditional physics instruction and in the Karlsruhe approach]. *Physica Didactica, 18*(2/3), 57–75.

Kesidou, S., Duit, R., & Glynn, S. M. (1995). Conceptual development in physics: Students' understanding of heat. In S. Glynn & R. Duit (Eds.), *Learning science in the schools* (pp. 179–198). Mahwah, NJ: Lawrence Erlbaum.

Kubli, F. (2005). Science teaching as a dialogue: Bakhtin, Vygotsky and some applications in the classroom. *Science & Education, 10*, 595–599.

Lancor, R. (2014a). Using student-generated analogies to investigate conceptions of energy. A multidisciplinary study. *International Journal of Science Education, 34*, 1–23.

Lancor, R. (2014b). Using metaphor theory to examine conceptions of energy in biology, chemistry, and physics. *Science & Education, 23*, 1245–1267.

Martin, B. E., & Brower, W. (1991). The sharing of personal science and the narrative element in science education. *Science Education, 75*(6), 707–722.

Niebert, K., Marsch, S., & Treagust, D. F. (2012). Understanding needs embodiment: A theory-guided reanalysis of the role of metaphors and analogies in understanding science. *Science Education, 96*(5), 849–877.

Starauschek, E. (2001). *Physikunterricht nach dem Karlsruher Physikkurs* [Physics instruction according to the Karlsruhe Approach]. Berlin: LOGOS.

Treagust, D. F., Duit, R., Joslin, P., & Lindauer, I. (1992). Science teachers' use of analogies: Observations from classroom practice. *International Journal of Science Education, 14*, 413–422.

Treagust, D. F., & Tsui, C. Y. (Eds.). (2013). *Multiple representations in biology education*. Dordrecht: Springer.

Treagust, D. F., Venville, G. J., Harrison, A. G., & Dagher, Z. (1996). Using an analogical teaching approach to engender for interpreting conceptual change. *International Journal of Science Education, 18*, 213–229.

Tobin, K. (1990). Changing metaphors and beliefs: A master switch for teaching. *Theory into Practice, 29*(2), 122–127.

Vollmer, G. (1984). Mesocosm and objective knowledge. In M. Wuketits (Ed.), *Concepts and approaches in evolutionary epistemology* (pp. 69–121). Dordrecht: Reidel Publishing Company.

Wilbers, J., & Duit, R. (2006). Post-festum and heuristic analogies. In P. Aubusson, A. Harrison, & S. M. Ritchie (Eds.), *Metaphor and analogy in science education* (pp. 37–49). Dordrecht: Springer.

Yore, L., & Treagust, D. F. (2006). Current realities and future possibilities: Language and science literacy—Empowering research and informing instruction. *International Journal of Science Education, 28*, 291–314.

# Conceptual Metaphor and the Study of Conceptual Change: Research synthesis and future directions

Tamer G. Amin

*Science and Mathematics Education Center, Department of Education, American University of Beirut, Beirut, Lebanon*

Many of the goals of research on conceptual metaphor in science education overlap with the goals of research on conceptual change. The relevance of a conceptual metaphor perspective to the study of conceptual change has already been discussed. However, a substantial body of literature on conceptual metaphor in science education has now emerged. This work has not yet been synthesized or related explicitly to the goals of conceptual change research. This paper first presents a broad sketch of the study of conceptual change, characterizing the goals of this body of work, its contributions to date, and identifying open questions. Next, the literature on conceptual metaphor in science education is reviewed against this background. The review clarifies the natural theoretical connections between the conceptual metaphor perspective and the phenomenon of conceptual change. It then examines the contributions made by the literature on conceptual metaphor in science education to the goals of research on conceptual change—namely, characterizing student conceptions, identifying obstacles to learning, understanding the process of conceptual change, and designing productive pedagogical strategies that could achieve conceptual change. The paper concludes with a discussion of further avenues for research into conceptual change, suggested by adopting a conceptual metaphor perspective.

## Introduction

This paper reviews the contributions of a growing literature on conceptual metaphor in science education to the study of conceptual change in science learning and instruction. There is a vast literature on conceptual change in science education and related fields. This work has characterized learners' pre-instruction conceptions and expert scientists' conceptual understanding, contrasted learners' initial conceptions with those of scientists, proposed accounts of the process of concept learning and applied the emerging understanding to the design of curricula and instructional environments (for reviews see Amin, Smith, & Wiser, 2014; diSessa, 2006; Duit & Treagust, 2003). In an early paper, Andersson (1986) used the cognitive linguistic theory of conceptual metaphor (Lakoff & Johnson, 1980, 1999) to identify a common feature of learner preconceptions in many domains of science (what he called an 'experiential gestalt of causation'). More recently, a body of literature has emerged that is applying this theory to a wide range of issues in science education (e.g. Amin, 2009; Amin, Jeppsson, Haglund, & Strömdahl, 2012; Brookes & Etkina, 2007, 2009; Gupta, Hammer, & Redish, 2010; Jeppsson, Haglund, Amin, & Strömdahl, 2013; Lancor, 2013, 2014a; Niebert, Marsch, & Treagust, 2012; Scherr et al., 2013; Scherr, Close, McKagan, & Vokos, 2012). This research has recognized that implicit in the language of science are systematic metaphorical mappings between abstract scientific concepts—such as *heat*, *energy*, and *entropy*—and concrete image schemas such as *material object/substance*, *possession*, *containment*, *object movement*, and *forced object movement*.[1] These implicit mappings (referred to as conceptual metaphors) are reflected in the language of science, as in 'The molecule *has* kinetic energy'; 'The energy *stored in* the compression of the spring was released'; and 'Heat was *lost* to the surroundings'. This literature has been exploring the implications of this phenomenon for science learning and instruction.

In Amin (2009), I analyzed the conceptual metaphors implicit in everyday English and in the language of science that are used to construe the concept of energy. I used this analysis to identify how image schemas (abstractions from sensorimotor experience) such as possession, containment, movement of possessions, and forced movement of possessions are used to construe energy in lay and scientific contexts. I argued that identifying these image schemas helps science educators identify continuity across the learning process. That is, a learner can draw on image schemas he or she already has as a cognitive resource to learn an abstract scientific concept like energy. I also suggested that naïve preconceptions might originate in construals implicit in everyday language (many of them metaphorical). But I also suggested that the implicitly metaphorical language of science can, itself, cue productive resources for the learner. So while I acknowledged that overly literal interpretations of metaphorical language might be a source of naïve preconceptions, I hypothesized that appropriating conceptual metaphors implicit in language might be a source of conceptual change. Moreover, I proposed that conceptual metaphor analysis of scientific language might help in the design of visual representations that would support meaningful learning.

A few years on and there is now a substantial body of literature on conceptual metaphor in science education that is providing some support for these hypotheses and has been exploring other implications of the phenomenon of conceptual metaphor for science teaching and learning. The literature has documented that the phenomenon of conceptual metaphor in the language of science textbooks is pervasive and systematic (Amin, 2009; Amin et al., 2012). It has suggested that expert and novice reasoning and problem-solving rely on the coordination of conceptual metaphors and other cognitive resources (Dreyfus et al., 2014; Dreyfus, Gupta & Redish, 2015; Jeppsson, Haglund, Amin, & Strömdahl, 2013). It has compared how novices and experts use conceptual metaphors in scientific problem-solving (Jeppsson, Haglund, and Amin, 2015). Moreover, it has shown that the construct of conceptual metaphor is useful for characterizing student conceptions (Lancor, 2014a, 2014b, 2015) and for identifying the source of naïve conceptions (Brookes & Etkina, 2007, 2009, 2015). It has used the perspective to evaluate the effectiveness of instructional analogies (Amin et al., 2012; Niebert et al., 2012) and to design instructional interventions, based on the strategic design of representations, that would develop meaningful understanding of challenging scientific concepts (Brewe, 2011; Close & Scherr, 2015; Niebert & Gropengießer, 2015; Scherr et al., 2013).

It is clear that many of the goals of this work on conceptual metaphor in science education overlap with the goals of research on conceptual change. The purpose of the present paper is to synthesize the substantial body of literature on conceptual metaphor in science education, discuss its contributions to the study of conceptual change and identify directions for future work. To do this, the paper first presents a broad sketch of research on conceptual change. Next, the literature on conceptual metaphor in science education is reviewed against this background. The review clarifies the natural theoretical connections between the conceptual metaphor perspective and the phenomenon of conceptual change. It also examines the contributions made by the literature on conceptual metaphor in science education to the goals of research on conceptual change—namely, characterizing student and scientist conceptions, identifying obstacles to learning, understanding the process of conceptual change, and designing productive instructional environments that could achieve conceptual change. In a final section, I discuss what further avenues for research into conceptual change can be opened up if we take a conceptual metaphor perspective, but point out that the perspective itself may need a more elaborated account of concepts.

## A Thumbnail Sketch of Research on Conceptual Change

In this section, I present a highly condensed sketch of research on conceptual change, relying primarily on a recent historical review of this literature spanning the last four to five decades (Amin et al., 2014). This sketch offers a perspective on the conceptual change literature with the specific purpose of clarifying the emerging contributions of research on conceptual metaphor in science education.

Conceptual change research has been conducted from a very wide range of perspectives and, as with any healthy scientific endeavor, disagreements between researchers persist (see Vosniadou, 2013a, for contributions representing a wide range of perspectives on conceptual change). However, Amin et al. (2014) argue that across this diversity of views, we can discern three phases of a trajectory of progress. In the 1970s and 1980s, the dominance of Piaget's stage view of development gave way to a domain-specific view of conceptual development and learning (Carey, 1985). Researchers recognized that it is necessary to describe changes in the content and structure of the learner's prior conceptions in order to make sense of how a learner comes to understand a scientific concept in some domain (e.g. force and motion; heat, temperature and energy; living things). Many detailed qualitative descriptions of learners' conceptions prior to, and during, instruction in a domain were reported in the literature. Learner conceptions were found to differ in significant ways from scientists' conceptions and were seen as obstacles to, and yet important starting points for, successful instruction. Creating cognitive conflict by making explicit and then challenging learners' naïve conceptions was proposed as an effective way to initiate instruction for conceptual change (see Driver & Easley, 1978; Scott, Asoko, & Driver, 1992, for reviews).

During the 1980s and 1990s, researchers scrutinized these conceptions and their transformations closely and attempted to induce change through instruction that targeted specific problematic conceptions. This scrutiny uncovered various components of the process of conceptual change. Four components were most widely studied. One component was the role of beliefs and assumptions about broad classes of entities (ontological categories) within which concepts were classified (e.g. it was proposed that many scientific concepts such as heat, energy, and electric current are incorrectly classified as material substances) (e.g. Chi, Slotta, & De Leeuw, 1994). The second component was the role of metacognitive beliefs *about* knowledge, learning, and science, such as *to learn is to remember* and *scientists arrive at new knowledge through discovery* (e.g. Hofer & Pintrich, 1997). The third was the role of useful intuitions and concrete/familiar conceptual structures that could serve as analogs of scientific concepts. These were strategically invoked by providing models or guiding students through modeling activities (e.g. the intuitive understanding of the agency of a compressed spring can be recruited to understand the concept of the normal force exerted by an apparently inert object such as a table) (Brown & Clement, 1989). Finally, the fourth component was the role of social interaction through which conflicting views trigger concept revision (e.g. Howe, Tolmie, & Rodgers, 1992) and collective thought supports the construction of more sophisticated knowledge (e.g. Hatano & Inagaki, 1991).

A third phase of research now sees researchers embracing the need to understand conceptual change as a complex process with multiple components and interactions. Some researchers realized early on the importance of understanding the complexity of learners' conceptual 'ecologies' and the multiple knowledge types and interactions that influence learner conceptions and the process of change (e.g. diSessa, 1993; Strike & Posner, 1985). A broader consensus on this point (not always explicitly acknowledged as such) now seems to be emerging (Brown & Hammer, 2008; diSessa, 2002; Vosniadou, 2013b; Wiser & Smith, 2013). Moreover, the design of instructional interventions and curricula has also begun to take this complexity seriously (e.g. Corcoran, Mosher, & Rogat, 2009).

Just as there is now widespread recognition that many types of knowledge play an important role in conceptual change, understanding the *representational format* of that knowledge is also seen as significant (Carey, 2009; Cheng & Brown, 2010;

diSessa, 1993). Researchers contrast *iconic* and *propositional* representations, albeit not always using this terminology. Iconic representations are analogical representations that bear a resemblance to what they represent such as imagery, image schemata, and mental models. Imagery is the mental reenactment, or simulation, of a previous perceptual experience in the absence of the object or events—for example, imagining rods, coils, and springs of different forms in an effort to predict whether a weight stretches springs with different diameter coils to different degrees (Clement, 1989). Image schemata are abstractions from sensorimotor experiences and are invoked to dynamically and causally interpret perceived or imagined objects and events—for example, what diSessa (1993) has called p-prims, such as *force-as-mover*. Together imagery and image schemata support the construction of mental models that can serve as analogs of physical objects and events. Researchers have shown that mental models support learners' and scientists' creative insights during scientific reasoning and problem-solving (Clement, 2009). Propositional representations are composed of arbitrary, symbolic representations that are constructed according to formal rules and express a claim about the world (i.e. they have a 'truth value'). Examples include linguistic expressions such as *A whale is a mammal* and mathematical representations, such as $F = ma$.

Brown (1993) contrasts these types of representations in terms of the degree to which they are explicit and accessible to conscious thought. Image schemata (or what Brown calls 'core intuitions') are triggered automatically in particular contexts and most likely remain implicit, beyond conscious awareness. Propositional ('verbal–symbolic') representations are by nature explicitly invoked and guide conscious chains of reasoning. Imagery and mental models can be either implicit or explicit. Making these distinctions has been important in understanding concept development and learning. During the normal course of conceptual development some concepts develop by simply assembling and jointly invoking iconic representations (e.g. lay concept of *animal*), while others require integrating iconic and propositional representations (e.g. the concept of natural numbers) (see Carey, 2009; Mandler, 2004). During science learning, conceptual understanding, and reasoning of students drawing on integrated iconic and propositional representations are more powerful than those who rely on iconic representations alone (Cheng & Brown, 2010).

I assume that what has been sketched thus far is either agreed on by most conceptual change researchers or tacitly assumed. However, theoretical diversity and points of disagreement are important to acknowledge. Theoretical diversity can be seen in the 'grain size' seen to be important in the characterization of learner and scientist conceptual understanding (diSessa, 2006). In order of *decreasing* grain size, researchers have characterized conceptual understanding in terms of (unanalyzed) theories—for example, a naïve impetus theory (McCloskey, 1983); mental models—for example, a naïve source–recipient model of heat (Wiser, 1995); framework theories analyzed in terms of ontological presuppositions—for example, the presupposition that unsupported things fall which constrains the construction of naïve models of the earth, and the beliefs generated from them (Vosniadou & Brewer, 1992); and image schematic phenomenological primitives (p-prims)—for example, *force-as-mover* (diSessa, 1993).

Related to this contrast in preferences for grain size, researchers have disagreed on two key (related) points: the extent to which pre-instruction conceptions are likely to be coherent and stable; and the extent to which conceptual change should be viewed as restructuring as opposed to gradually increasing organization of knowledge elements triggered strategically in specific contexts. A broad range of views can be identified where conceptual change is viewed as ontological recategorization (arguably the most extreme coherence view) (Chi, 2005; Chi et al., 1994); revision of conceptual structures embedded in framework theories with an emphasis on the revision of ontological presuppositions (Vosniadou, 2013b); coordinated revision of networks of beliefs (domain specific and epistemological) and mental models (Wiser & Smith, 2013); and the gradual organization of multiple (often intuitive) cognitive resources, which assumes that pre-instruction knowledge is highly fragmented (diSessa, 2002). While these views are usually presented in the literature as in opposition to one another, Brown and Hammer (2008) have suggested that most can be reinterpreted as special cases of a more general account of conceptions and conceptual change formulated in terms of dynamic systems theory.

Among the open questions that are only just beginning to be explored are the following: How can the coherence and fragmentation of conceptions be studied within a single unifying perspective that can accommodate both phenomena on a case-by-case basis (e.g. Brown & Hammer, 2008)? What are the multiple knowledge elements that need to be integrated for successful conceptual change to be achieved (e.g. diSessa, 2014; Wiser & Smith, 2013)? More specifically, how are propositional language-like representations integrated with iconic knowledge structures during conceptual change (e.g. Cheng & Brown, 2010; Jeppsson et al., 2013; Sherin, 2001, 2006)? How can instruction and curriculum design guide the assembly of multiple knowledge elements of various types (Corcoran et al., 2009)?

## Investigating Conceptual Metaphor in Science Education: Taking stock of contributions to the study of conceptual change

In this section, I review the literature on conceptual metaphor in science education in light of the sketch of the conceptual change literature just presented. The goal is to clarify the contributions that this body of work makes to the study of conceptual change. The review first highlights natural theoretical connections between the cognitive linguistic theory of conceptual metaphor and the phenomenon of conceptual change. Next, I present the contributions of research on conceptual metaphor in science education to characterizing student and scientist conceptions and identifying obstacles to learning, understanding the process of conceptual change, and designing productive instructional strategies that could achieve conceptual change.

*Natural Theoretical Connections Between Conceptual Metaphor and Conceptual Change*

There are natural theoretical connections between the cognitive linguistic theory of conceptual metaphor and conceptual change. As we saw in the sketch above, at the heart of research on conceptual change is the question of how pre-instruction conceptions are transformed to increasingly approximate conceptions sanctioned by scientists. Of particular interest is how pre-instruction conceptions of the novice learner contribute to the conceptual structures of the (emerging) expert scientist. The theory of conceptual metaphor has a ready response to these questions, albeit at a general level.

The central claim of the theory of conceptual metaphor is that abstract concepts are understood metaphorically in terms of more concrete knowledge structures (Lakoff & Johnson, 1980, 1999). That is, concrete source domains are mapped metaphorically onto abstract target domains; the mapping facilitates understanding and reasoning in the abstract domain by drawing on the intuitive understanding of the concrete source domain. I use the descriptor 'concrete' to refer to the fact that many concepts (e.g. chairs, sitting, and dog) can be represented directly in terms of iconic representations such as image schemas, which are generalizations over sensorimotor experiences. In contrast, I use 'abstract' to describe concepts that cannot be represented directly in terms of such perception and motor-based experience. Abstract concepts will need to be represented in terms of propositional representations such as language or mathematical symbols. For the latter to be understood in a way that goes beyond pure manipulation of symbols, metaphorical projection from image schematic knowledge structures would be needed. For example, our conception of time cannot simply be described in terms of perception-based imagery or generalizations over sensorimotor experiences. Language, numbers, and other representations are needed. But making sense of these representations relies on mappings from spatial conceptions that are based on perception and action. That is, recurring patterns in our everyday sensorimotor experiences result in more general knowledge structures with multiple related components, gestalts, such as *moving objects, paths with a starting point and destination*, and *obstacles to movement along a path*. These gestalts—so-called image schemas—are mapped onto the concept of time. The result is that time is construed metaphorically in terms of image schemas and inferences inherited from the structure of these spatial image schemas (see Lakoff & Johnson, 1999, for extended discussion of the metaphorical understanding of time). For example, we speak of 'approaching' a deadline; getting to some important point in a career as being 'a long journey'; and of having to 'get things out of the way' before beginning a new project. These metaphorical uses reflect systematic underlying conceptual mappings such as A Moment In Time Is A Location Along A Path; Passage Of Time Is Movement Toward A Location; and Carrying Out Intermediate Tasks Is Removal Of Obstacles. Inferences that we would arrive at intuitively in the spatial domain—where objects move along a path and removing obstacles can be needed to reach a destination—map onto and support inferences in the domain of time. From a conceptual metaphor perspective, developing an understanding of the concept of time is (at least in part) to construct the appropriate mapping between spatial image schemas and the conceptual domain of time (see Williams, 2011/2012, for an investigation into how a teacher guides children into constructing these mappings as they learn to tell the time).

Thus, the theory of conceptual metaphor is ready with a partial answer to the question: How do learners understand abstract concepts in terms of resources already available to them? That is, expert scientific understanding can be understood (again, at least in part) as the strategic use of image schematic knowledge structures to construe abstract concepts metaphorically. Sometimes understanding an abstract scientific concept may require multiple metaphorical mappings, with different conceptual metaphors used in different contexts. Therefore, to understand a concept, the learner will need to construct the appropriate mappings and draw strategically on a number of conceptual metaphors across contexts. From this perspective, a misconception could result from drawing on an inappropriate source domain in a particular context or incorrectly mapping an aspect of an appropriate source domain onto the abstract target.

The grounding of understanding of abstract scientific concepts in generalizations from sensorimotor experience has been recognized in the science education literature for some time in the constructs of phenomenological primitives (diSessa, 1993) and anchoring intuitions (Clement, 1993) and in model-based instruction designed to trigger physical intuitions (White, 1995). The theory of conceptual metaphor suggests that analyzing language can help us identify the image schemas that ground abstract scientific understanding. It also raises questions about how learners might be able to establish the appropriate metaphorical mappings sanctioned by science and points to instructional interventions that might support learning. The following section illustrates how these themes have been examined in research on conceptual metaphor in scientific expertise, science learning, and instruction.

## Contributions to the Goals of Conceptual Change Research

In this subsection, I review the literature on conceptual metaphor in science education, highlighting how this literature contributes to the same goals of research on conceptual change—namely, characterizing scientist and learner conceptions and identifying obstacles to learning; understanding the process of conceptual change; and suggesting productive pedagogical strategies.

*Characterizing learner and scientist conceptions and identifying obstacles to learning.* As described earlier, a key claim of the theory of conceptual metaphor is that in human cognition concrete source domains are frequently mapped onto abstract domains and that these mappings are reflected in metaphorical expressions. This is a mundane phenomenon, not a feature of special creative thought. Thus, we should expect that this phenomenon is a common feature of the thinking of scientists and learners and will be frequently reflected in scientific and lay language. A number of studies have used the conceptual metaphor framework to describe scientist and learner conceptions and to suggest possible obstacles to learning by comparing the two.

Most of these studies have adopted a similar approach. They begin by analyzing scientists' language use from a conceptual metaphor perspective. This analysis identifies how abstract concepts in some scientific domain are construed in terms of image schemas. They then identify student conceptions in this domain and interpret these conceptions from a conceptual metaphor perspective. The findings from the two analyses are then compared with an eye to answering some or all of the following questions: (1) How do the source domains used by scientists and students to ground their understanding of some domain compare? (2) Do scientists and learners differ in the contexts in which they invoke particular source domains to metaphorically construe a scientific concept? (3) When the same source is selected to construe the same abstract concept, does the mapping differ?

In Amin (2009), I presented an analysis of the use of the term energy in *The Feynman Lectures on Physics* from a conceptual metaphor perspective. I described what conceptual metaphors are used by Feynman to construe various aspects of the concept of energy—transport, transformation, and conservation (degradation was not addressed). Multiple metaphorical mappings were used to construe each aspect of the concept of energy, but a great deal of systematicity could be discerned. For example, the many metaphorical construals of energy transport involved construing energy as a substance/possession and energy transport as movement of a possession between components of a physical system; energy transformation was also construed in terms of the movement of a possession, with energy often construed as a stored resource. In contrast, forms of energy themselves were construed as containers. Construing energy as a substance, as movement on a vertical scale and in terms of a part–whole schema helped in making sense of energy conservation.

This analysis of a scientific text was followed by a review of the literature on learners' alternative conceptions of energy. Then to help in interpreting these conceptions, the use of the term energy in everyday English was analyzed from a conceptual metaphor perspective. In that analysis, it was found that reference to energy came up in the context of human use of technology, food consumption, and human activity and vitality. As in the scientific text, there was also pervasive use of metaphor in everyday English. Again there was construal of energy as a location on a vertical scale, as a resource stored or contained and changes in energy as movement of a substance. Most of learners' alternative conceptions of energy found in the science education literature corresponded to metaphorical construals that are implicit in everyday language. On comparing the two sets of mappings (in Feynman's text and those in everyday English), a great deal of overlap was noted in which source domains were used to construe energy. One key difference was the absence of the part–whole schema as a source domain in everyday English, which was associated only with energy conservation in the scientific text.

Brookes and Etkina (2007, 2009) took a similar approach in the domains of quantum mechanics, and force and motion. Again, as a first step, they identified metaphorical expressions in scientists' language and inferred the underlying conceptual mappings between more concrete conceptual domains and abstract domains. They then studied student reasoning and characterized their conceptions in terms of the mapping of concrete onto abstract domains. In this work, they point out that the scientists' language implies a particular ontological categorization of concepts (Chi et al., 1994) and examine how this language is interpreted by learners. For example, Brookes and Etkina (2007) point out that the scientific language of quantum mechanics includes implicit metaphors in which a potential well is construed as a physical container. In a later study (Brookes & Etkina, 2009), they show that in the history of scientific thinking about force, four distinct metaphors were explored, most of which continue to find a place in the modern language of science: force as an agent, force as an internal drive, force as a passive medium of interaction and force as a property of an object. In both studies, they provide evidence that student misconceptions mirror these ontological metaphors implicit in the language they are exposed to during instruction. Similarly, in a recent study (Brookes & Etkina, 2015), they have shown that university students reason about heat incorrectly in some contexts as if it is a state function, mirroring metaphorical language inviting a construal of heat as a substance. In all these cases, the source of learner misconceptions seems to be that they map too much of the source domain implied by scientists' language.

Lancor (2013, 2014a, 2014b, 2015) has used the tools of conceptual metaphor to develop a framework to analyze how university students understand and communicate their ideas about the scientific concept of energy in a variety of contexts. In Lancor (2014a), students were asked to write analogies that would explain their understanding of the role of energy in ecosystems (biology), chemical reactions (chemistry), and mechanical systems and electric circuits (physics). She classified student conceptions in terms of a framework composed of six conceptual metaphors: energy as a substance that can be accounted for; energy as a substance that can change forms; energy as a substance that can be carried; energy as a substance that can be lost; energy as a substance that can be an ingredient or product; and energy as a process or interaction. Lancor found that students used these conceptual metaphors differently in different disciplinary contexts. In Lancor (2015), she uses the same analytical framework to analyze student essays explaining the role of energy in interdisciplinary scientific contexts widely discussed in the media, including radiation, transportation, generating electricity, earthquakes, and the big bang theory. Across these two studies, Lancor finds that the same framework can be used to analyze students' conceptions in disciplinary and interdisciplinary contexts. Both within, and across studies, different conceptual metaphors are favored in different contexts.

Niebert and Gropengießer (2015) also use the conceptual metaphor perspective to analyze learner conceptions. In addition, they analyze scientist conceptions and compare the two. They suggest that analyzing conceptions in terms of conceptual metaphors is a particularly useful 'grain size'. It allows researchers to generalize across conceptions without losing out on interesting variation. It also helps describe the differences between scientist and learner conceptions, enabling the researcher/

educator to specify the 'learning demand' for a particular scientific topic. Niebert and Gropengießer are particularly interested in the bodily basis of the image schematic source domains of the conceptual metaphors they examine. They adopt the view that human beings are designed to make sense of their world at the level of medium-scale dimensions, what they call the mesocosm (in between the micro- and macrocosm). This is the scale at which the senses evolved to function. Sense making in science (by both learner and scientist) usually involves the attempt to draw on image schemas to make sense of more abstract ideas associated with the microcosm (e.g. cell division) or macrocosm (e.g. carbon cycle).

Niebert and Gropengießer use a conceptual metaphor perspective to identify the difficulties that learners can face in a specific scientific domain. For example, learners might map only a *container* image schema onto the target concept of myelin sheath making it difficult for them to make sense of the effect of myelin on signal speed in neurons. Scientists, in contrast, use the image schema of a *bridge* together with the *container* schema to construe the myelin sheath. This enables them to conceptualize signal travel as *jumping*, which helps them understand the effect of myelin on signal speed. Niebert and Gropengießer generalize across the different domains in which they analyze learner and scientist conceptions. They suggest that they need to appeal to a limited number of image schemas to understand student and scientist conceptions and that students and scientists often make use of the same ones but in different ways. They argue that the difficulties students have are often with selecting which ones to map to which target concepts, but note that sometimes student difficulties are due to the absence of the relevant experiences needed to construct the right image schema.

The literature on conceptual metaphor in science education reviewed in this subsection makes a number of contributions to the study of conceptual change. It provides an additional tool to help in characterizing student conceptions, with an emphasis on what image schemas students use to understand a scientific concept and how they use them. The conceptual metaphor framework also suggests a way of analyzing how *scientific* concepts are understood. In much of the conceptual change literature, the researcher assumes knowledge of the scientific concept that is the target of learning and instruction. Adopting a conceptual metaphor perspective encourages the analyst to ask what image schemas are used to understand a scientific concept. In turn, identifying the image schemas that both students and scientists use in a particular scientific context and comparing the two, we can identify a learning demand in the context. This type of analysis contributes to the broader goal that some conceptual change researchers see as important of identifying the intuitive knowledge structures relevant to learning in some domain and describing the learning challenge in terms of these structures (e.g. Brown, 1993; Brown & Hammer, 2008; Clement, 2009; diSessa, 1993, 2014).

*Understanding the process of conceptual change.* A number of conclusions can be drawn from the literature reviewed in the previous subsection that are relevant to attempts to understand the *process* of conceptual change. First, even before instruction, learners metaphorically map image schematic source domains unconsciously onto target conceptual domains. This mapping is relevant to instruction, but the relevance can be either positive or negative. On the positive side, some mappings are similar to those implicit in scientific language and thought and so contribute positively to learning. On the negative side, some either involve mapping the wrong source domain or involve an incorrect mapping of the right source domain, potentially leading to misconceptions in both cases.[2] Second, the research reviewed earlier has shown that many of the source domains drawn on by learners and scientists consist of image schematic knowledge structures, which are abstractions from sensorimotor experience. Finally, the close connection between these cross-domain mappings and language implicates language in the process of conceptual change itself. I take each of these points in turn, discussing implications for understanding the process of conceptual change.

Since the phenomenon of conceptual metaphor is a mundane characteristic of human cognition, science learners have already implicitly constructed a number of mappings by the time they begin to receive formal instruction in some topic. Some of the mappings they have constructed are consistent with those in the scientific domain they are learning, but others are not. Those that are not consistent will sometimes be the source of misconceptions and so will need to be revised during the concept learning process. One could then say that *an aspect* of conceptual change is the revision of metaphorical mappings between source and target domains. If this claim is correct, this has implications for the debate in the conceptual change literature regarding the extent to which conceptual change involves transformation of a fairly coherent naïve conceptual structure to another coherent (more) scientific conceptual structure. As reviewed earlier, researchers have disagreed on the degree of coherence of learner pre-instruction conceptions. Given that the phenomenon of conceptual metaphor is a fairly systematic phenomenon, even in everyday language and thought, some degree of systematicity in the mappings implicit in learner thinking is to be expected. This of course is something that needs to be investigated on a case-by-case basis in each domain of interest. One example of such coherence deriving from metaphorical mappings in learner thinking is the ontological classification implicit in the language of science in the domains of quantum mechanics and force and motion, investigated by Brookes and Etkina (2007, 2009); another is the pervasive use of the energy as substance metaphor in everyday language and thought (Amin, 2009; Vosniadou, 2009).

Many of the source domains of the conceptual metaphors found in the thought of learners and scientists are image schematic knowledge structures such as possession, movement of possession, movement along a path, and containment. These are abstractions from sensorimotor experience and are constructed early in life through repeated interactions with the physical world. A conceptual metaphor perspective suggests that learning a scientific concept involves learning to map particular image schemas to abstract scientific concepts strategically. Learners need to map the right image schemas in the right way to the abstract scientific concepts they are trying to understand. Assuming that most image schemas originate in sensorimotor experiences early in life implies continuity across the learning process at the level of image schemas. This claim is equivalent to the claim made in the conceptual change literature that an important aspect of learning scientific concepts is the reorganization of phenomenological primitives (p-prims) (diSessa, 1993). Indeed, many p-prims can be seen as image schematic source domains of conceptual metaphors (see Amin, 2009, and Jeppsson et al., 2013, for discussion). Jeppsson et al. (2013) have suggested that scientists use image schemas strategically to construe scientific concepts while solving problems. For example, they found that construing heat metaphorically as a substance and a mathematical function as a machine that takes substance as input helps in seamlessly coordinating qualitative and quantitative reasoning. This strategic use of image schemas seems to be an aspect of what needs to be learned when expertise is acquired. Jeppsson et al. (2015) have recently suggested that novices do not display the same productive use of image schemas when engaged with the same problems, but are often limited to invoking conventional metaphorical expressions that they have been exposed to in pedagogical discourse.

Finally, conceptual metaphors are reflected in language—that is, in metaphorical *expressions*. This implicates language in the process of conceptual change. Pre-instruction learner conceptions have been found to mirror conceptual metaphors reflected in everyday language (Amin, 2009). Moreover, ontological misclassifications by learners on initial exposure to a scientific domain have been traced to conceptual metaphors implicit in the language of science (Brookes & Etkina, 2007, 2009, 2015). Moreover, the pervasiveness and systematicity of the phenomenon of conceptual metaphor in everyday language and in the language of science textbooks is well documented (e.g. Amin, 2009; Amin et al. 2012). This means that both informal and formal concept acquisition are likely to involve the appropriation of metaphorical construals implicit in language. The cognitive developmental literature has recognized a role for language in the process of conceptual development in specific areas (Carey, 2009; Gentner, 2010). These considerations suggest that language (through the conceptual metaphors that it expresses) may have a significant influence on conceptual change in the context of science learning (see Amin, 2009, for discussion). However, much more direct empirical evidence is needed to support this claim.

*Suggesting productive pedagogical strategies.* A conceptual metaphor perspective on student and scientist conceptions and on the process of conceptual change has pedagogical implications. Some of these have only been hypothesized as implications of empirical studies of scientific expertise and learning, while others have been empirically investigated directly.

Taking a conceptual metaphor perspective has suggested a tool for the formative assessment of student conceptions. Lancor (2013) has suggested that a framework of seven conceptual metaphors can be used as a formative assessment tool to provide useful information about student conceptions of energy. The framework clarifies what each conceptual metaphor affords in understanding aspects of the concept of energy and what its limitations are. Thus, the framework allows teachers to identify some resources students have for thinking about energy and to anticipate difficulties they might face. Lancor suggests that this form of assessment is preferrable to directly asking students 'What is energy?'

I showed earlier that the conceptual metaphor perspective can be used to analyze scientific language to reveal how scientists make use of image schemas to conceptualize abstract scientific concepts. Uncovering the image schematic grounding of a concept can guide the design of visual representations that are likely to support learning by strategically triggering the appropriate image schemas (Amin, 2009). This is consistent with Mathewson's (2005) case for the important role of 'master images' in scientific understanding. Mathewson argues that across a wide range of scientific domains key images such as conduits, containers, paths, boundaries, and others recur to ground understanding of abstract concepts. Conceptual metaphor analysis can help identify master images that are implicit in the language of some scientific domain.

The imagistic basis of understanding scientific concepts has prompted a number of researchers to design visual representations to help students understand challenging concepts. Scherr and colleagues have developed a program of research, The Energy Project, that uses a conceptual metaphor perspective to design representations, and instructional environments more generally, to improve understanding of the concept of energy (Close & Scherr, 2015; Scherr, Close, Close, & Vokos 2012; Scherr, Close, McKagan et al., 2012, 2013). To design the instructional environment, they begin by assuming that energy is construed metaphorically as substance-like and that this construal is central to a scientific understanding of energy transfer and transformation. Based on this assumption, Scherr, Close, McKagan et al. (2012) evaluated common representations of energy and revealed their strengths and limitations. They pointed out that common representations such as bar charts, pie charts, and others do not adequately afford conceptualizing energy as a substance-like entity that flows between components of a system. In follow-up research, Scherr, Close, Close et al. (2012) designed energy tracking representations to support modeling of energy flow and transformation in physical processes. In addition, they have designed the Energy Theater learning environment where participants use their bodies to represent units of energy and regions on the floor marked with rope represent components of a physical system (Close & Scherr, 2015; Scherr et al., 2013). The Energy Theater was designed to help in-service teachers participating in a professional development program develop their understanding of the concept of energy. It encouraged teachers to engage in conceptually rich discussions about how to model energy flow and transformation and what representations best capture aspects of the concept of energy. When these discussions were analyzed it was evident that teachers' understanding of the concept of energy improved by participating in this professional development program.

Assuming that energy is construed metaphorically as a substance in scientific thinking has also prompted a *curricular* innovation for teaching introductory university physics (Brewe, 2011). In designing this curriculum, Brewe selects modeling as an important aspect of scientific thinking for learners to develop and a useful vehicle for concept learning. Brewe suggests that when energy is conceptualized as a substance, energy transfer, storage, and transformation are easily represented visually and these representations facilitate modeling of physical systems. In his curriculum, the concept of energy is introduced before the concept of force (in a reversal from traditional instruction) and students return to the topic of energy repeatedly throughout the course. This approach allows students to make connections across topics that are typically presented in isolation. While this novel curricular approach has not been systematically evaluated, Brewe presents a qualitative analysis of classroom discussion as evidence for its promise.

Interest in the concept of energy has dominated much of the work applying a conceptual metaphor perspective to the design of instructional repesentations and learning environments in science education. However, Niebert and Gropengieβer (2015) have used this perspective to design instructional representations in other conceptual domains. They have addressed the topics of cell division, the greenhouse effect, nerve signal transmission, and the carbon cycle. In this work, they use a conceptual metaphor perspective to analyze student and scientist conceptions so as to characterize the learning demand in each domain. Based on this analysis, they design external representations that they use in teaching experiments. They illustrate that they are able to use these representations to get learners to map an image schema that they had previously failed to map to a particular target scientific concept.

The conceptual metaphor perspective has also been used to evaluate instructional analogies. Amin et al. (2012) analyzed the conceptual metaphors implicit in pedagogical discourse dealing with entropy and the second law of thermodynamics. They identified which conceptual metaphors are commonly and systematically used in a range of science textbooks to construe entropy and the second law. They argued that this can help educators select instructional analogies that are likely to be particularly effective, if it is assumed that learners must appropriate the mappings implicit in the language of science they encounter. If this assumption is correct, then an explicit instructional analogy is more likely to be effective if the mappings it involves are consistent with those implicit in the language students encounter. For example, Amin et al. hypothesized that entropy-as-freedom is likely to be more effective than entropy-as-disorder. They showed how the mappings of the entropy-as-freedom analogy are more consistent with the mappings of the conceptual metaphors implicit and yet pervasive in pedagogical discourse on the topic. Whether or not this analogy is particularly effective still needs to be confirmed empirically. Niebert et al. (2012) also used a conceptual metaphor perspective, but took another approach to evaluating instructional metaphors and analogies. In their study, they surveyed the science education literature and reanalyzed 199 instructional metaphors and analogies from a conceptual metaphor perspective. They concluded that effective metaphors and analogies are those that incorporate source domains that derive from bodily based experience—that is, image schematic source domains.

Amin et al. (2012) considered the pedagogical implications of conceptual metaphors implicit in the scientific language that students encounter. In contrast,

Haglund, Jeppsson, and Ahrenberg (2014) have examined the instructional impli-cations of conceptual metaphors implicit in *everyday* language. Using analyses of large everyday language corpora (in English and Swedish), Haglund et al. (2014) reveal the extensive metaphorical use of the word 'momentum' in many everyday con-texts including sports, politics, and others. They compare how the word is used in these everyday contexts to how it is used in physics. They conclude that there is sub-stantial overlap in these uses, in contrast to the much noted discrepancy between the everyday and scientific use of 'force'. They argue that their analysis suggests that it might be productive to introduce the concept of momentum in science curricula much earlier than is typical.

In sum, a conceptual metaphor perspective can guide the design of formative assess-ment tools and instructional representations and learning environments, can help in the evaluation and selection of analogies, and can suggest curricular innovations. The role of representations and analogies as tools to induce conceptual change is well known (e.g. Clement, 2009; Gentner, 2010; Gilbert & Treagust, 2009). However, the per-spective of conceptual metaphor offers a novel basis to approach the design and selec-tion of representations and analogies. That is, the design and selection can be peformed with knowledge of the image schematic grounding of scientific concepts and the meta-phors implicit in everyday and scientific language.

## New Directions for Research on Conceptual Change Suggested by Adopting a Conceptual Metaphor Perspective

We saw in the previous section that the existing literature on conceptual metaphor in science education makes a number of contributions to the study of conceptual change. However, this research has been presented so far as if it draws on a single, monolithic perspective. Moreover, a positive slant has been taken to point out the contributions of the perspective to studying conceptual change. In this section, I discuss points of con-trast between the assumptions made by different researchers in this emerging area of research as well as some limitations of the perspective. Highlighting these points of contrast and limitations raises empirical questions about the role of conceptual meta-phor in conceptual change and points to directions for further theoretical develop-ment. So in this section, I discuss the potential for further contributions suggested by adopting a conceptual metaphor perspective. Three themes are highlighted: the contrasting views of language, either as a tool for the researcher or as a tool for thought; stability versus contextual variation in conceptualization; and the need for an explicit account of concepts that incorporates the phenomenon of conceptual metaphor within it.

### Language as a Medium of Conceptualization Versus a Tool for the Researcher

As is clear from what has been discussed in this paper so far, there is a close connec-tion between language and conceptual metaphor. The systematic mappings between conceptual domains are reflected in metaphorical expressions. However, researchers using a conceptual metaphor perspective in science education have treated language differently. This difference can be expressed as the difference between viewing language as a tool for the researcher or as a medium of conceptualization and, thus, a tool of thought (Budwig, 1999). Many of the researchers reviewed earlier have treated language as a tool for the researcher to identify underlying metaphorical conceptions of students, teachers, or scientists (e.g. the work of Close, Lancor, Niebert, Scherr, and colleagues). In this work, when those studied use specific linguistic markers (e.g. verbs or prepositions), researchers treat these as indicating the use of image schematic knowledge structures to conceptualize physical situations. The idea that there are systematic mappings between particular abstract scientific concepts and certain image schemas is used to categorize instances of participant conceptions into broad classes (Lancor, 2014a, 2014b; Niebert & Gropengießer, 2015). Moreover, instructional analogies are evaluated by determining whether the source domains of the analogies consist of readily accessible image schemas (Niebert et al., 2012). In addition, when instructional environments have been designed in a way that embodies a conceptual metaphor (e.g. the construal of energy as a substance or vertical location), the focus has been on the design of *visual* representations (Scherr, Close, McKagan et al., 2012) and three-dimensional models (e.g. the human body in the Energy Theater strategy) (Close & Scherr, 2015; Scherr et al., 2013).

Using language as a tool for the researcher to identify student conceptions is reasonable (indeed necessary). However, it does not address the role of language as a tool of thought for the learner. Some of the work drawing on the construct of conceptual metaphor in science education has suggested that concept learning may involve appropriating language-based conceptual metaphors (Amin, 2009; Amin et al., 2012; Brookes & Etkina, 2007, 2009, 2015). In Amin (2009), I suggested that naïve conceptions of energy can be traced to metaphorical construals implicit in everyday language. I also suggested that metaphorical expressions in scientific language might trigger source domains that help learners understand scientific concepts. Brookes and Etkina (2007, 2009, 2015) have provided evidence for the importance of this process of appropriation. They show that overly concrete interpretations of metaphorical expressions in the language of science can lead to misconceptions. By showing that learning obstacles can be traced to construals implicit in the language of science, they are providing evidence that appropriation of construals implicit in scientific language is part of the concept learning process.

This finding leads to an important conclusion: metaphorical expressions invite the reader/listener to map some source domain onto some target, but the details of the mapping may vary from one person to another. Considering language as a tool for the learner raises various questions: How are learners' and scientists' interpretations of metaphorical expressions in the language of science different? What features of prior knowledge and the design of learning environments guide learners to interpret such expressions as scientists do? Moreover, an important methodological question arises when considering language as a tool for the researcher: How can a researcher infer the nature of student conceptions when a piece of linguistic data is observed? Further research that addresses these questions is needed.

*Stability Versus Contextual Variation in Conceptualization*

The stability versus the contextual variation in conceptualization has been debated in the conceptual change literature for some time. A clear example of this debate has been the disagreement over the role of ontological classification of concepts in conceptual change. Chi (2005) and Chi et al. (1994) have argued that many misconceptions arise from a misclassification of scientific concepts within a substance, as opposed to a constraint-based interaction, ontological category. On this view, concept learning is seen to involve the reclassification of concepts from one incorrect, yet stable, ontological view to another. diSessa (1993) has argued that pre-instruction conceptions are not stable in this way, and questions the extent to which ontological classification of a concept can provide an adequate picture of pre-instruction knowledge.

In the science education literature using a conceptual metaphor perspective, parallel contrasting views on ontology in learner conceptions can be identified. In their initial work from this perspective, Brookes and Etkina (2007) assimilated the phenomenon of conceptual metaphor and the misconceptions that arise from interpretations of metaphorical language into Chi's ontological view (2005; Chi et al., 1994). They argued that students' overly literal interpretations of substance metaphors implicit in the language of science lead them to ontological misclassification of some scientific concepts. In contrast, Amin (2001) suggested that ontological construal of heat in student thinking could possibly *emerge* in specific reasoning contexts. Moreover, Dreyfus et al. (2014) and Gupta et al. (2010) have appealed to the phenomenon of conceptual metaphor to critique Chi's view. They argue that the pervasive use of substance metaphors by learners as well as scientists suggests a more flexible, dynamic view of ontological classification of concepts. Gupta, Elby, and Conlin (2014) provide evidence for very productive use of a substance construal of gravity by teachers in a professional development program while trying to explain why objects of different masses fall with the same acceleration. Similarly, Jeppsson et al. (2013) provide evidence of flexible ontological construals in the context of scientific problem-solving. More recently, Brookes and Etkina (2009, 2015) have acknowledged the variation in ontological construals of concepts across contexts of use (although noting some stability *within* contexts). The stable ontologies proponents tend to advocate avoiding ontologically misleading language during instruction. In contrast, those granting flexibility and the productive use of 'naïve' ontologies argue that this instructional strategy is not practical given the pervasive use of implicit metaphor in scientific language and could be harmful because it suppresses potentially useful conceptions. There is empirical evidence supporting both views. So before instructional implications of ontologies implicit in language can be adequately explored some resolution of this debate is needed.

It has been suggested (Amin et al., 2014; Jeppsson et al., 2013) that one way out is to make the distinction between explicit ontological stances and implicit ontological construals. Stability might be discerned when someone is asked to explicitly state his or her ontological belief regarding some concept or to reason with a concept when the ontology is explicitly represented. On the other hand, the phenomenon of pervasive use of conceptual metaphors implicit in language includes subtle shifts in construal that users of these metaphors are not aware of. Thus, it may be theoretically coherent to acknowledge *both* explicit conceptual stability and implicit contextual variability. Another approach to this dilemma might be to accept that the issue of ontological stability versus contextual variability may be resolved differently in different cases. For example, there might be stability in some cases due to consistency in the use of a conceptual metaphor (e.g. the substance metaphor of energy) (Amin, 2009; Vosniadou, 2009). A third possibility is to acknowledge that multiple ontologies might be used but combined into a stable conceptualization. This latter possibility is suggested by the work of Dreyfus et al. (2015) who show that both a learner and scientist readily blend two ontological metaphors: energy-as-substance and energy-as-location. Repeated use of this blend in certain contexts (e.g. reasoning about physical systems when considering a potential energy graph) might stabilize this blend of two ontological construals. This third possibility has become theoretically visible through Dreyfus et al.'s use of the theory of conceptual blending (Fauconnier & Turner, 2002) as an analytical framework. All three possibilities are theoretically plausible phenomena. Further empirical research addressing the question of ontological stability versus contextual flexibility is needed that considers all three hypotheses.

*Incorporating Conceptual Metaphor into a View of Concepts*

Throughout this paper, the word 'concept' has been used but there has been no explicit discussion of what the term means. Indeed, there is no explicit discussion of how to understand concepts in the literature on conceptual metaphor in science education. Moreover, the previous two subsections reveal that two themes—the relationship between language and conceptualization, and between stable and contextually varying conceptions—need greater clarity in the literature on conceptual metaphor in science education (if not the literature on conceptual change more generally). This clarity is needed if a sustained program of research is to emerge. In this final subsection, I briefly, yet explicitly suggest a way to incorporate the phenomenon of conceptual metaphor into a view of concepts that could help establish this emerging literature on conceptual metaphor in science education as a programmatic effort.

There already are various theoretical perspectives on concepts and conceptual change in the science education literature, with two perspectives dominating over the last two to three decades: the 'coherence' (e.g. Vosniadou, 2013b; Wiser & Smith, 2013) and 'knowledge-in-pieces' (e.g. diSessa, 1993, 2002, 2014) perspectives. The literature drawing on the notion of conceptual metaphor reviewed in this paper suggests that at least this construct needs to be incorporated into any account. But neither the coherence nor the knowledge-in-pieces views have made much progress in incorporating attention to the role of language in concept representation and change (see Amin, 2009, for a discussion). Finally, the fault line of stability versus contextual variation no longer seems to reflect an accurate representation of disagreements in the science education literature on conceptual change more broadly (Amin et al., 2014). For all these reasons, attempting to formulate a new perspective seems warranted.

It is proposed here that a version of the view of concepts put forward by Susan Carey (2009) can help to address these issues. Very briefly, in Carey's view, concepts per se are understood as unitary, language-like symbols, whereas the *content* of a concept is understood in terms of the network of inferences in which it participates (its 'inferential role') and how it refers to entities in the world. The inferential role of a concept is specified in terms of a network of beliefs expressed in terms of propositional representations (e.g. language and mathematical representations) in which the concept participates as well as the iconic knowledge structures (e.g. images, image schemas, and mental models) that interpret these propositions.

I will illustrate this view of concepts by using the case of *energy*. Let us accept that the concept per se is simply some mental token triggered by the word energy (or symbol E). In order to characterize the content of this concept we must characterize the network of beliefs that it participates in (e.g. 'Energy is conserved across physical transformations'; 'Energy is transformed *from* potential *to* kinetic energy as an object falls to the ground'; 'When an object hits the ground, energy is *exchanged between* the object and the ground'; $KE = \frac{1}{2}mv^2$).[3] These statements are interpreted and generate further beliefs based on images and image schemas, both possibly assembled into larger scale analogical mental models. These propositionally expressed beliefs and the imagistic representations that interpret them constitute part of the concept's inferential role. Those situations in the physical world to which it is appropriate to apply these statements and serve as the basis for their contextual interpretation constitute the second aspect of the content of the concept, its referential component. Carey does not discuss conceptual metaphor, but this phenomenon can be incorporated into this account through the propositionally expressed beliefs that implicitly mark mapping between the abstract concept of interest (e.g. energy) and image schemas. The linguistic pointers to these mappings are italicized in the examples above—the conceptual metaphors illustrated being Forms Of Energy Are Containers and Change Of Energetic State Is Transfer Of A Possession/Substance. Thus, the image schemas of *container* and *transfer of possession/substance* will contribute to the inferential role of the concept.

While this presentation is brief, I hope it can give a sense of how this view of concepts takes the role of propositional (language and mathematics) and iconic representations (images, image schemas, and mental models) seriously. Distinguishing these knowledge types and acknowledging their joint contributions to conceptual understanding in science education is not new (see Brown, 1993; diSessa, 1993), but this has not been discussed in an explicit account of concepts per se. Moreover, this brief discussion has pointed out where the construct of conceptual metaphor relates to the notion of concept.

This view does not commit to one side of the longstanding stability versus contextual variation debate or the other. From this perspective a scientific concept's inferential role is made up of a vast network of beliefs applied adequately across a wide range of physical situations. Conceptual stability can be represented via the repeated (and joint) use of subcomponents of the network of beliefs/image schemas, whereas contextual variation can be captured by inconsistent and independent use of these subcomponents (Wiser & Smith, 2013). The extent of stability/contextual variation for both students and scientists should be seen as an empirical question examined in each domain of interest (see also Brown & Hammer, 2008, for an account in which stability and contextual variation can both be accommodated within a dynamic systems perspective). Finally, this account changes our view of ontology and how it relates to conceptual metaphor. The content of a concept is represented in terms of a network of propositional and iconic representations, with conceptual metaphors (sometimes concrete substance metaphors) implicit in beliefs. Therefore, an instance of use of metaphorical expression that reflects a conceptual metaphor (e.g. energy-as-substance) cannot be seen as classification *of the concept* within an ontological category, but rather a local, momentary construal. Instead, how concepts might be classified within broad ontological categories has to be determined by comparing whole networks of beliefs with one another.

This view of concepts suggests questions for future research. Two themes are highlighted here: first, characterizing student conceptions; and second, evaluating and designing instructional interventions that focus on the use of instructional analogies. First, with regard to characterizing student conceptions, the approach taken so far has been to catalogue the implicit metaphors in student speech and then infer the conceptions they hold. The research by Lancor reviewed earlier exemplifies this point. Identifying substance language to infer substance construals of energy in various scientific contexts is useful but constitutes a small part of characterizing students' *concepts* of energy. These metaphorical conceptions are implicit in beliefs that are themselves part of larger networks which need to be described. Moreover, the scope of application of these beliefs to physical situations needs to be determined. Future research from a conceptual metaphor perspective will need to conduct these more comprehensive characterizations. Second, with regard to evaluating instructional analogies, it is important to keep in mind that the content of concepts is to a large extent represented in terms of propositionally expressed beliefs. While these beliefs are *interpreted* in terms of images, image schemas, and models, these iconic representations do not fully characterize the target concept. Therefore, evaluating instructional analogies purely in terms of the extent to which accessible image schematic source domains are employed or strategically triggered (see e.g. Niebert & Gropengießer, 2015) is not sufficient. It is important to also evaluate the propositionally expressed beliefs that they encourage as well as their scope of application.

The extent to which the proposal made here will help synthesize and motivate programmatic research on conceptual metaphor in science education can only be evaluated in hindsight. Any framework can only be evaluated by how productive it turns out to be. The goal of this proposal was to give a sense of key features of a view of concepts that incorporates the construct of conceptual metaphor and the types of questions that it motivates.

## Conclusions

This paper has tried to show that a literature on conceptual metaphor in science education is emerging with goals that overlap significantly with those of research on conceptual change. This work has identified image schemas that students invoke when trying to understand scientific concepts; these can be productive but can also lead to misconceptions. It has identified multiple image schemas that ground understanding of scientific concepts through metaphor. It has shown that implicit metaphorical mapping between domains can be described by analyzing lay and scientific language. Moreover, it suggests that an aspect of the process of conceptual change might be the appropriation of metaphorical construals implicit in language. These findings have instructional implications. They suggest a way to approach formative assessment of student conceptions by identifying their repertoire of conceptual metaphors for a given concept. They suggest ways to design instructional representations that would trigger productive image schemas and select promising analogies that are consistent with conceptual metaphors implicit in pedagogical discourse. In addition, they have curricular implications, such as pointing to productive points of entry in some domain.

Identifying points of difference among researchers working on conceptual metaphor in science education suggests some directions for future work. In this paper, I have discussed differences in how language is viewed, either as a tool for the researcher or as a tool for thought. The fact that interpreting metaphorical language can vary from person to person poses challenges to the researcher trying to infer student conceptions from linguistic data. It also raises questions about what conditions (student knowledge and instructional environments) encourage interpretations that are more consistent with those of scientists. I have also discussed differences in assumptions about the stability versus contextual variability in conceptualization, with a focus on ontological classification. Different researchers provide evidence pointing to both stability and variability in ontological classification of concepts by learners and scientists. A number of hypotheses (not mutually exclusive) were put forward that might resolve this disagreement. I suggested that the degree of stability may vary from case to case; that it might be necessary to distinguish explicit ontological classification and implicit construals, where the former is stable and the latter contextually variable; and that stable and yet combined use of more than one ontological construal is a third resolution. Further work is needed to decide if any or all of these cases are plausible and common. Finally, the literature on conceptual metaphor in science education lacks, at present, an explicit view of the nature of concepts and one that incorporates the phenomenon of conceptual metaphor. I described one such view where the content of a concept is characterized in terms of a network of beliefs (the concept's inferential role) and the situations in the world to which it refers. This network of beliefs will include statements that incorporate conceptual metaphors. This way of thinking about concepts suggests a number of directions for future work using a conceptual metaphor perspective. Chief among them is the need to describe learners' use and interpretation of metaphors as part of a wider characterization of their network of beliefs in a conceptual domain.

## Acknowledgements

I would like to thank David Brown and Rachel Scherr for very helpful comments on an earlier version of this paper.

## Disclosure statement

No potential conflict of interest was reported by the author.

## Notes

1. The use of the descriptors "abstract" and "concrete" in this paper will be clarified later.
2. This latter point regarding metaphorical mappings implicit in everyday language is parallel to the point made in the science education literature about explicit instructional analogies. Even if a potentially productive source domain is provided by teachers, incorrect mappings performed by students can lead to misconceptions (see Glynn, 1989).
3. In principle, such a network is limitless! However, space limitations prohibit engaging with this issue here. See Carey (2009) for a proposal for how to identify that subset of beliefs that is key to the characterization of the content of a concept.

## References

Amin, T. G. (2001). A cognitive linguistics approach to the layperson's understanding of thermal phenomena. In A. Cienki, B. Luka, & M. Smith (Eds.), *Conceptual and discourse factors in linguistic structure* (pp. 27–44). Stanford, CA: CSLI.

Amin, T. G. (2009). Conceptual metaphor meets conceptual change. *Human Development, 52*(3), 165–197.

Amin, T. G., Jeppsson, F., Haglund, J., & Strömdahl, H. (2012). Arrow of time: Metaphorical construals of entropy and the second law of thermodynamics. *Science Education, 96*(5), 818–848.

Amin, T. G., Smith, C., & Wiser, M. (2014). Student conceptions and conceptual change: Three overlapping phases of research. In N. Lederman & S. Abell (Eds.), *Handbook of research in science education, vol II* (pp. 57–81). New York, NY: Taylor and Francis.

Andersson, B. (1986). The experiential gestalt of causation: A common core to pupils' preconceptions in science. *European Journal of Science Education, 8*(2), 155–171.

Brewe, E. (2011). Energy as a substance-like quantity that flows: Theoretical considerations and pedagogical consequences. *Physical Review Special Topics: Physics Education Research, 7*(2), 0201006, 1–14.

Brookes, D. T., & Etkina, E. (2007). Using conceptual metaphor and functional grammar to explore how language used in physics affects student learning. *Physical Review Special Topics: Physics Education Research, 3*(1), 010105.

Brookes, D. T., & Etkina, E. (2009). 'Force,' ontology and language. *Physical Review Special Topics: Physics Education Research, 5*(1), 010110, 1–13.

Brookes, D. T., & Etkina, E. (2015). The importance of language in students' reasoning about heat in thermodynamics processes. *International Journal of Science Education.* doi:10.1080/09500693.2015.1025246

Brown, D. E. (1993). Refocusing core intuitions: A conceretizing role for analogy in conceptual change. *Journal of Research in Science Teaching, 30*(10), 1273–1290.

Brown, D., & Clement, J. (1989). Overcoming misconceptions via analogical reasoning: Abstract transfer versus explanatory model construction. *Instructional Science, 18*(4), 237–261.

Brown, D., & Hammer, D. (2008). Conceptual change in physics. In S. Vosniadou (Ed.), *International handbook of research on conceptual change* (pp. 127–154). New York, NY: Routledge.

Budwig, N. (1999). The contribution of language to the study of the mind: A tool for researchers and children. *Human Development, 42,* 362–368.

Carey, S. (1985). Are children fundamentally different types of thinkers and learners than adults? In S. Chipman, J. Segal, & R. Glaser (Eds.), *Thinking and learning skills* (Vol. 2, pp. 485–517). Hillsdale, NJ: Erlbaum. Reprinted by Open University Press: Open University Readings in Cognitive Development.

Carey, S. (2009). *The origin of concepts.* New York, NY: Oxford University Press.

Cheng, M. F., & Brown, D. E. (2010). Conceptual resources in self-developed explanatory models: The importance of integrating conscious and intuitive knowledge. *International Journal of Science Education, 32*(17), 2367–2392.

Chi, M. T. H. (2005). Common sense conceptions of emergent processes. *The Journal of the Learning Sciences, 14,* 161–199.

Chi, M. T. H., Slotta, J. D., & De Leeuw, N. (1994). From things to processes: A theory of conceptual change for learning science concepts. *Learning & Instruction, 4*(1), 27–43.

Clement, J. (1993). Using bridging analogies and anchoring intuitions to deal with students' preconceptions in physics. *Journal of Research in Science Teaching, 30*(10), 1241–1257.

Clement, J. (1989). Learning via model construction and criticism: Protocol evidence on sources of creativity in science. In J. A. Glover, R. R. Ronning, & C. R. Reynolds (Eds.), *Handbook of creativity* (pp. 341–381). New York, NY: Plenum Press.

Clement, J. (2009). *Creative model construction in scientists and students: The role of imagery, analogy, and mental simulation.* Dordrecht: Springer.

Close, H., & Scherr, R. (2015). Enacting conceptual metaphor through blending: Learning activities embodying the substance metaphor for energy. *International Journal of Science Education.* doi:10.1080/09500693.2015.1025307

Corcoran, T., Mosher, F., & Rogat, A. (2009). *Learning progressions in science: An evidence-based approach to reform* (Unpublished CPRE Research Report #RR-63). Teachers College, Columbia University.

diSessa, A. (2002). Why 'conceptual ecology' is a good idea. In M. Limon & L. Mason (Eds.), *Reconsidering conceptual change. Issues in theory and practice* (pp. 29–60). Dordrecht: Kluwer Academic.

diSessa, A. (2006). A history of conceptual change research: Threads and fault lines. In R. Sawyer (Eds.), *The Cambridge handbook of the learning sciences* (pp. 265–281). Cambridge, MA: Cambridge University Press.

diSessa, A. A. (1993). Toward an epistemology of physics. *Cognition and Instruction, 10*(2–3), 105–225.

diSessa, A. A. (2014). The construction of causal schemes: Learning mechanisms at the knowledge level. *Cognitive Science, 38*(5), 795–850.

Dreyfus, B. W., Geller, B. D., Gouvea, J., Sawtelle, V., Turpen, C., & Redish, E. F. (2014). Ontological metaphors for negative energy in an interdisciplinary context. *Physical Review Special Topics: Physics Education Research, 10,* 020108.

Dreyfus, B. W., Gupta, A. & Redish, J. (2015). Applying conceptual blending to model coordinated use of multiple ontological metaphors. *International Journal of Science Education*. doi:10.1080/09500693.2015.1025306

Driver, R. & Easley, J. (1978). Pupils and paradigms: A review of the literature related to concept development in adolescent science students. *Studies in Science Education*, 5, 61–84.

Duit, R., & Treagust, D. F. (2003). Conceptual change: A powerful framework for improving science teaching and learning. *International Journal of Science Education*, 25(6), 671–688.

Fauconnier, G., & Turner, M. (2002). *The way we think: Conceptual blending and the mind's hidden complexities*. New York, NY: Basic Books.

Gentner, D. (2010). Bootstrapping the mind: Analogical processes and symbol systems. *Cognitive Science*, 34(5), 752–775.

Gilbert, J. K., & Treagust, D. F. (2009). *Multiple representations in chemical education*. Dordrecht: Springer.

Glynn, S. M. (1989). The teaching with analogies model. In K. D. Muth (Ed.), *Children's comprehension of text: Research into practice* (pp. 185–204). Newark, NJ: International Reading Association.

Gupta, A., Elby, A., & Conlin, L. D. (2014). How substance-based ontologies for gravity can be productive: A case study. *Physical Review Special Topics: Physics Education Research*, 10, 010113.

Gupta, A., Hammer, D., & Redish, E. F. (2010). The case for dynamic models of learners' ontologies in physics. *Journal of the Learning Sciences*, 19(3), 285–321.

Haglund, J., Jeppsson, F., & Ahrenberg, L. (2014). Taking advantage of the 'Big Mo'—Momentum in everyday English and Swedish and in physics teaching. *Research in Science Education*. doi:10.1007/s11165-014-9426-x

Hatano, G., & Inagaki, K. (1991). Sharing cognition through collective comprehension activity. In L. B. Resnick, J. M. Levine, & S. D. Teasley (Eds.), *Perspectives on social shared cognition* (pp. 331–348). Washington, DC: American Psychological Association.

Hofer, B., & Pintrich, P. (1997). The development of epistemological theories: Beliefs about knowledge and knowledge their relations to learning. *Review of Educational Research*, 67, 88–140.

Howe, C., Tolmie, A., & Rodgers, C. (1992). The acquisition of conceptual knowledge in science by primary school children: Group interaction and the understanding of motion down an incline. *British Journal of Developmental Psychology*, 10, 113–130.

Jeppsson, F., Haglund, J., & Amin, T. (2015). Varying use of conceptual metaphor across levels of expertise in thermodynamics. *International Journal of Science Education*. doi:10.1080/09500693.2015.1025247

Jeppsson, F., Haglund, J., Amin, T. G., & Strömdahl, H. (2013). Exploring the use of conceptual metaphor in solving problems on entropy. *Journal of the Learning Sciences*, 22(1), 70–120.

Lakoff, G., & Johnson, M. (1980). *Metaphors we live by*. Chicago, MA: University of Chicago Press.

Lakoff, G., & Johnson, M. (1999). *Philosophy in the flesh*. New York, NY: Basic Books.

Lancor, R. A. (2013). The many metaphors of energy: Using analogies as a formative assessment tool. *Journal of College Science Teaching*, 42(3), 38–45.

Lancor, R. A. (2014a). Using student-generated analogies to investigate conceptions of energy: A multidisciplinar study. *International Journal of Science Education*, 36(1), 1–23.

Lancor, R. A. (2014b). Using metaphor theory to examine conceptions of energy in biology, chemistry, and physics. *Science & Education*, 23(6), 1245–1267.

Lancor, R. A. (2015). An analysis used by students to describe energy in an interdisciplinary general science course. *International Journal of Science Education*. doi:10.1080/09500693.2015.1025309

Mandler, J. (2004). *The foundations of mind: The origins of conceptual thought*. Oxford, MA: Oxford University Press.

Mathewson, J. H. (2005). The visual core of science. *International Journal of Science Education, 27,* 529–548.

McCloskey, M. (1983). Naïve theories of motion. In D. Gentner & A. Stevens (Eds.), *Mental models* (pp. 299–324). Hillsdale, NJ.: Erlbaum.

Niebert, K., & Gropengießer, H. (2015). Understanding starts in the mesocosm: Conceptual Metaphor as a Framework to develop external representations for science teaching. *International Journal of Science Education.* doi:10.1080/09500693.2015.1025310

Niebert, K., Marsch, S., & Treagust, D. F. (2012). Understanding needs embodiment: A theory-guided reanalysis of the role of metaphors and analogies in understanding science. *Science Education, 96*(5), 849–877.

Scherr, R. E., Close, H. G., Close, E. W., Flood, V. J., McKagan, S. B., Robertson, A. D., & Vokos, S. (2013). Negotiating energy dynamics through embodied action in a materially structured environment. *Physical Review Special Topics: Physics Education Research, 9*(2), 020105.

Scherr, R. E., Close, H. G., Close, E. W., & Vokos, S. (2012). Representing energy. II. Energy tracking representations. *Physics Review Special Topics: Physics Education Research, 8*(2), 020115 1–11.

Scherr, R. E., Close, H. G., McKagan, S. B., & Vokos, S. (2012). Representing energy. I. Representing a substance ontology for energy. *Physical Review Special Topics: Physics Education Research, 8*(2), 020114.

Scott, P. H., Asoko, H., & Driver, R. H. (1992). Teaching for conceptual change: Review of strategies. In R. Duit, F. Goldberg, & H. Niederer (Eds.), *Research in physics learning: Theoretical issues and empirical studies* (pp. 310–329). Kiel: IPN - Institute for Science Education.

Sherin, B. (2001). How students understand physics equations. *Cognition and Instruction, 19*(4), 479–541.

Sherin, B. (2006). Common sense clarified: The role of intuitive knowledge in physics problem-solving. *Journal of Research in Science Teaching, 43*(6), 535–555.

Strike, K. A., & Posner, G. J. (1985). A conceptual change view of learning and understanding. In L. H. West & A. L. Pines (Eds.), *Conceptual structure and conceptual change* (pp. 189–210). Orlando: Academic Press.

Vosniadou, S. (2009). Yes to embodiment, no fragmentation: Commentary on Amin 2009. *Human Development, 52*(3), 198–204.

Vosniadou, S. (2013a). *International handbook of research on conceptual change* (2nd ed.). New York, NY: Routledge.

Vosniadou, S. (2013b). Conceptual change in learning and instruction: The framework theory approach. In S. Vosniadou (Ed.), *International handbook of research on conceptual change* (2nd ed. pp. 11–30). New York, NY: Routledge.

Vosniadou, S., & Brewer, W. F. (1992). Mental models of the earth: A study of conceptual change in childhood. *Cognitive Psychology, 24,* 535–585.

White, B. Y. (1995). The ThinkerTools Project: Computer microworlds as conceptual tools for facilitating scientific inquiry. In S. M. Glynn & R. Duit (Eds.), *Learning science in the schools: Research reforming practice* (pp. 201–227). Hillsdale, NJ: Lawrence Erlbaum Associates.

Williams, R. F. (2011/2012). Image schemas in clock-reading: Latent errors and emerging expertise. *Journal of the Learning Sciences, 21*(2), 216–246.

Wiser, M. (1995). Use of history of science to understand and remedy students' misconceptions about heat and temperature. In D. N. Perkins, J. L. Schwartz, M. M. West, & M. S. Stone (Eds.), *Software goes to school* (pp. 23–38). New York, NY: Oxford University Press.

Wiser, M., & Smith, C. (2013). Learning and teaching about matter in the middle school years: How can the atomic-molecular theory be meaningfully introduced? In S. Vosniadou (Ed.), *International handbook of research on conceptual change* (2nd ed. pp. 177–194). New York, NY: Routledge.

# Index

Note: Page numbers in **bold** type refer to figures
Page numbers in *italic* type refer to tables
Page numbers followed by 'n' refer to notes